Mathematics Teacher Education

Volume 11

Scope

The Mathematics Teacher Education book series presents relevant research and innovative international developments with respect to the preparation and professional development of mathematics teachers. A better understanding of teachers' cognitions as well as knowledge about effective models for preservice and inservice teacher education is fundamental for mathematics education at the primary, secondary and tertiary level in the various contexts and cultures across the world. Therefore, considerable research is needed to understand what facilitates and impedes mathematics teachers' professional learning. The series aims to provide a significant resource for teachers, teacher educators and graduate students by introducing and critically refl ecting new ideas, concepts and findings of research in teacher education.

More information about this series at http://www.springer.com/series/6327

Timo Leuders • Kathleen Philipp • Juliane Leuders
Editors

Diagnostic Competence of Mathematics Teachers

Unpacking a Complex Construct in Teacher Education and Teacher Practice

Springer

Editors
Timo Leuders
University of Education Freiburg
Freiburg, Germany

Juliane Leuders
University of Education Freiburg
Freiburg, Germany

Kathleen Philipp
School of Education
University of Applied Sciences and Arts
of Northwestern Switzerland
Basel, Switzerland

Mathematics Teacher Education
ISBN 978-3-319-66325-8 ISBN 978-3-319-66327-2 (eBook)
DOI 10.1007/978-3-319-66327-2

Library of Congress Control Number: 2017955370

Printed on acid-free paper

This Springer imprint is published by Springer Nature
The registered company is Springer International Publishing AG
The registered company address is: Gewerbestrasse 11, 6330 Cham, Switzerland

Contents

Part II Developing Diagnostic Competence and Applying Diagnostic Competence in the Classroom: Models for Teacher Training

Part I
The Nature of Diagnostic Competence: Conceptualizing and Measuring Facets of Diagnostic Competence

Diagnostic Competence of Mathematics Teachers: Unpacking a Complex Construct

Timo Leuders, Tobias Dörfler, Juliane Leuders, and Kathleen Philipp

Although diagnostic competence of teachers is regarded as a key component of successful teaching, there are many open questions regarding the structure, the development and the impact of diagnostic competence. This chapter presents an overview of different approaches to pinpoint diagnostic competence theoretically and to investigate it empirically: measuring judgment accuracy, assessing competences in diagnostic situations or analyzing judgment processes. These approaches are discussed with respect to their advantages, restrictions as well as some of their main findings and they are allocated within an overarching model of diagnostic competence as a continuum, comprising diagnostic dispositions, diagnostic thinking and diagnostic performance.

1 Diagnostic Competence: A Relevant but Complex Construct

Among the many tasks that teachers have to accomplish in everyday teaching, *diagnostic activities* are considered to be pivotal (Artelt & Gräsel, 2009; Berliner, 1994; Bromme, 1997; Darling-Hammond, 2000; Demaray & Elliott, 1998; KMK, 2014; Helmke, 2010; Hoge & Coladarci, 1989; Schrader, 2011; Weinert, 1998).

T. Leuders (✉) • J. Leuders
University of Education Freiburg, Freiburg, Germany
e-mail: leuders@ph-freiburg.de

T. Dörfler
University of Education Heidelberg, Heidelberg, Germany

K. Philipp
School of Education, University of Applied Sciences and Arts
of Northwestern Switzerland, Basel, Switzerland

T. Leuders et al. (eds.), *Diagnostic Competence of Mathematics Teachers*,
Mathematics Teacher Education 11, DOI 10.1007/978-3-319-66327-2_1

Diagnostic activities comprise the gathering and interpretation of information on the learning conditions, the learning process or the learning outcome, either by formal testing, by observation, by evaluating students' writings or by conducting interviews with students. The goal of diagnostic activities is to gain valid knowledge about individual students or the whole class in order to plan further individual support or whole class teaching, to inform students and parents or to decide on resources. The teachers' knowledge, skills, motivations and beliefs relevant to these diagnostic activities can be summarized as *diagnostic competences* (Aufschnaiter et al., 2015; Herppich et al., 2017; Schrader, 2011; Weinert, 2000). Under the umbrella term "diagnostic competence" research has been focusing on many different aspects, such as the accuracy of diagnostic judgments (*veridicality*, cf. Schrader & Helmke, 1987) or the knowledge and appropriate use of assessment procedures (*assessment literacy*, cf. Popham, 2009). In this sense, diagnostic competence cannot be regarded as *one* individual trait but rather as a broad construct bundle.

There is no doubt about the impact of diagnostic competences of teachers on the learning outcome of students. Experience from practice and findings from research on learning and teaching render the assumption highly plausible that diagnostic competence of teachers 'does matter'. However, its impact is mediated in a complex fashion by many aspects of the quality of teaching; for example, by taking into account previous knowledge of students, by applying adaptive teaching methods (e.g., Beck et al., 2008) or by giving formative feedback (e.g., Black & Wiliam, 2003). So, although we have some empirical evidence on the effect of diagnostic competences (Anders, Kunter, Brunner, Krauss, & Baumert, 2010; Carpenter, Fennema, Peterson, Chiang, & Loef, 1989; Schrader & Helmke, 1987), we still lack a clear understanding of the mechanisms of their impact on student learning. Many efforts to identify etiological factors of diagnostically substantial judgments and effective diagnostic teaching have not yet yielded a clear picture (see also Schrader, 2013; Südkamp, Kaiser, & Möller, 2012).

As an illustration of the challenges connected to the construct of diagnostic competence, consider the example in Fig. 1.

With this example, many questions arise, which imply different perspectives on the construct of diagnostic competence:

- What kind of conditions regarding the situation and the teacher allow her or him to recognize the information as diagnostically interesting and useful?
- What kind of knowledge is relevant to interpret the students' activity or response? What is the contribution of domain specific knowledge?
- In what way are the judgments triggered? What is the role of rational analysis and intuition, of explicit knowledge and implicit experience?
- How valid and how reliable are these interpretations? What are the implications of the uncertainty of diagnostic judgments?
- What are the consequences following the diagnostic judgment? What kind of decision is possible and justifiable? Which action would be inappropriate?
- Does this situation represent a typical and relevant diagnostic activity? Which other diagnostic activities arise in everyday teaching?

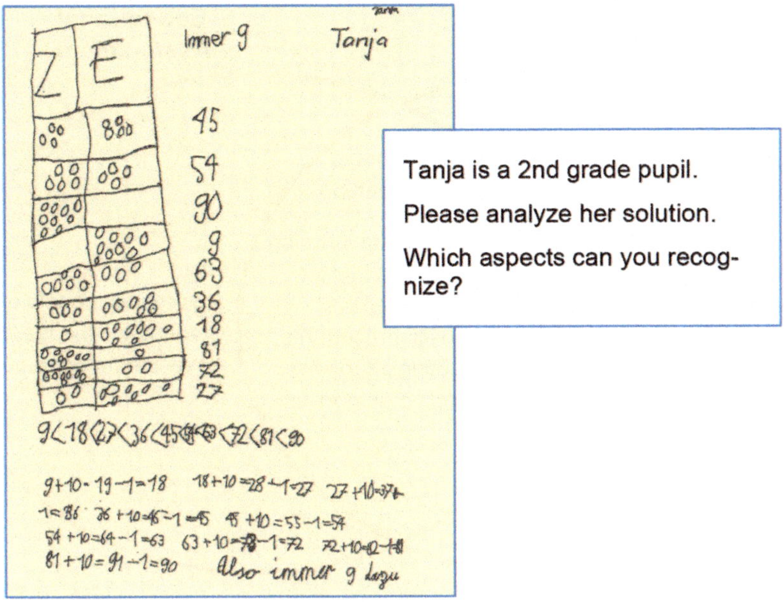

Fig. 1 A student's solution and the request for diagnostic interpretation (Hengartner, Hirt, & Wälti, 2007 p. 45, cited in Leuders & Leuders, 2013)

Interestingly, many of these questions are also discussed in the area of educational assessment, especially those of validity of diagnostic judgments (e.g., Messick, 1995). It is important that the notion of assessment is not restricted to the development, application and quantitative evaluation of formal tests but also comprises any classroom practice and any interpretive evaluation with the goal to procure information on students' traits and learning processes. So, informal diagnostic judgments of teachers are an important component of assessment. However, when we discuss the questions above from the perspective of diagnostic competence we have a strong interest in teachers' behavior and the traits and conditions that influence their behavior.

Furthermore, one should always bear in mind that diagnostic activities are only a subset of a teachers' activities. In practice they are closely linked to, for example, preparing teaching materials, guiding classroom activity, grading and giving feedback to students and parents among many others. For the sake of understanding teacher behavior and its conditions better, it is useful to analytically describe the underlying dispositions, the cognitive processes and the resultant behavior by a competence construct (Baumert & Kunter, 2006; Blömeke, Gustafsson, & Shavelson, 2015; Schoenfeld, 2010; Weinert, 2001). This approach is also applicable to the area of diagnostic competence and will be explained in detail in the next section.

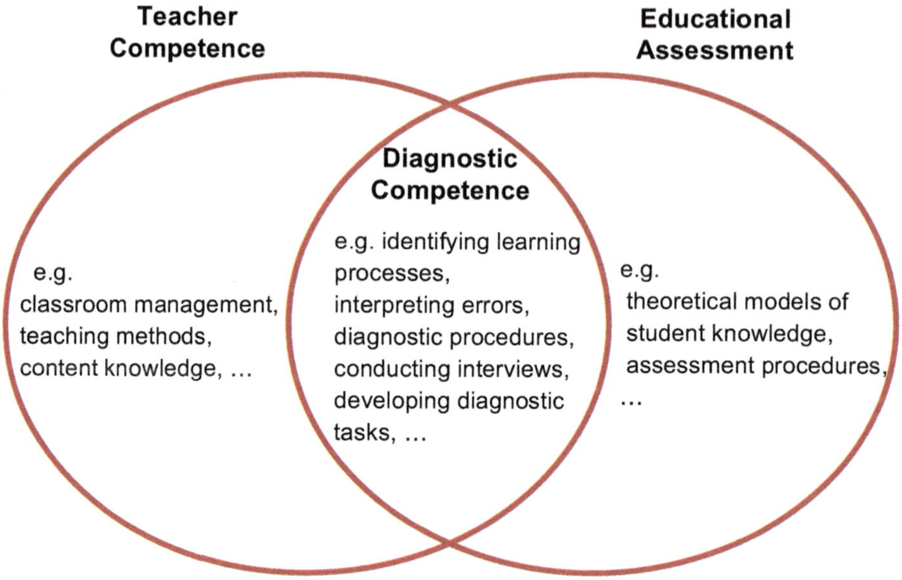

Fig. 2 Diagnostic competence as an area of interest in the intersection of the areas of teacher competence and educational assessment

Figure 2 illustrates the relation between the area of competence research on the one hand and assessment in education on the other, with diagnostic competence being located at the intersection of both.

The questions above, which arose in the context of the concrete example, all referred to the overarching research question: What is the development, the structure and the impact of the complex construct bundle called diagnostic competence? The type of research driven by this question certainly depends on the chosen theoretical framework. In the next sections, we give a synoptic overview of the frameworks currently used to investigate diagnostic competence.

1.1 What Is Diagnostic Competence? Terminological Considerations and a General Model

Although many authors use the term 'diagnostic competence', there is a wealth of understandings, of varying notions and of differing emphases with respect to its definition and constituents. One of the aims of this book is to collect and present a diversity of approaches related to diagnostic competence, but also to convey a unifying view on the construct that may help to discern important connections and to devise further measures in teacher education or teacher research. To adopt the term "diagnostic competence" instead of other frequently encountered terms, such as "diagnostic expertise", "diagnostic skills", "diagnostic abilities", "assessment skills", "assessment literacy", etc., has been a deliberate choice, which will be explained in the following.

It is well known that the use of terminology depends on the contingencies of traditions and cultures and of choices depending on communities of discourse. While in Anglo-American research on teacher education most authors prefer the term "knowledge" (e.g., Ball, Thames, & Phelps, 2008; Rowland & Ruthven, 2011; Shulman, 1986), in other contexts one increasingly often encounters the terms "competence" or "competency"[1], accompanied by a vivid theoretical discussion on the understanding and the significance of the underlying construct in general (Koeppen, Hartig, Kleine, & Leutner, 2008; Shavelson, 2010; Westera, 2001), and also with a special focus on higher education (Blömeke et al., 2015) and on teacher education (Baumert et al., 2010, Bromme, 1997).

Competences are considered to be "complex ability constructs that are context-specific, trainable, and closely related to real life" (Koeppen et al., 2008, p. 62). To be more specific, competences comprise "the readily available or learnable cognitive abilities and skills which are needed for solving problems as well as the associated motivational, volitional and social capabilities and skills which are required for successful and responsible problem solving in variable situations" (Weinert, 2001). The relation to effective behavior in a professional context is more explicit in the following definition (Spencer & Spencer, 1993, p. 9): "A competency is an underlying characteristic of an individual that is causally related to criterion-referenced effective and/or superior performance in a job or situation. *Underlying characteristic* means the competency is a fairly deep and enduring part of a person's personality and can predict behavior in a wide variety of situations and job tasks. *Causally related* means that a competency *causes* or *predicts* behavior and performance. *Criterion-referenced* means that the competency actually predicts who does something well or poorly, as measured on *a specific criterion* or standard".

When looking at these clarifications, it seems plausible to also use the notion of competence for dealing with the complex construct bundle connected to diagnostic activities. The following reasons support this choice:

- The situations in which diagnostic competence is applied are highly variable, complex real-life situations that bear a non-algorithmic, problem solving character (cf. the "teaching as problem-solving-perspective" (Schoenfeld, 2010)).
- Diagnostic competence certainly draws on knowledge, but also comprises skills (e.g., conducting interviews), as well as beliefs and affective and motivational components, such as a diagnostic stance (Hodgen, 2011; Prediger, 2010).

Blömeke et al. (2015) contend that research on competence since the 1970s – not restricted to the sector of general education, but also highly relevant in vocational education or medicine (clinical competence) – was following two competing strands: An *analytic approach* identifies relevant individual traits (cognitive,

[1] Some authors deliberately distinguish between competence (pl. competences) and competency (pl. competencies) (cf. Blömeke et al., 2015) but this fine distinction is not adopted for the purpose of the analyses in this book.

affective, motivational) and then potentially tries to develop training situations to foster these traits. In contrast to this, a *holistic approach* looks at certain realistic criterion situations, defines successful task performance in these situations by observable behavior and then potentially tries to identify (and recruit) individuals who are 'fit' for certain tasks.

There has been much research on teacher education during the last decades that has adopted the analytic approach, and tried to describe and identify relevant traits of teachers, especially knowledge dimensions such as pedagogical knowledge, subject-matter knowledge and pedagogical content knowledge. Within the holistic approach, one rather regards competences as unfolding dynamically within concrete situations (cf. Depaepe, Verschaffel, & Kelchtermans, 2013) and defines certain teacher behavior in classroom situations as an indicator of diagnostic competence.

However, the perspectives need not be mutually exclusive: Blömeke et al. (2015) propose a model of competence as a continuum, which embodies both perspectives and also includes cognitive processes that lead to observable behavior (Fig. 3).

When applying this model to the case of diagnostic competence, one recognizes that it can integrate three complementary perspectives on the construct. We specify this model with respect to the area of diagnostic activities and – by adopting the terminology of the authors – define the following areas:

- *Diagnostic dispositions* comprise knowledge, beliefs, motivational and affective factors that are relatively stable within a person, and which contribute to the ability to act successfully in diagnostic situations. Research on diagnostic competence that relies on this perspective strives to describe and identify latent traits by means of tests (written, video-based, etc.), which are considered to operationalize diagnostic situations.
- *Diagnostic skills* can be regarded as a set of situation-specific cognitive functions or processes of perception, interpretation and decision-making (one may as well use the term *diagnostic thinking*). On the one hand these processes draw on diagnostic dispositions, and on the other hand they lead to observable performance in diagnostic situations. Research on diagnostic competence that focuses on this perspective establishes and investigates models for cognitive processes that explain diagnostic judgments.
- *Diagnostic performance* relates to observable behavior in diagnostic situations as they arise in the professional life of a teacher. Research can, for example, investigate the teachers' assessment of their students' achievement or learning behavior, and their ensuing actions and then develop appropriate objective measures. Diagnostic performance can be regarded as the product of the diagnostic dispositions and diagnostic skills, as well as their operationalization.

This concept of "diagnostic competence as a continuum" provides a wide notion of competence, which comprises different approaches to understand and capture diagnostic competence. However, there remains a possible terminological irritation: Many authors actually designate the narrower area of dispositions with the term "competence" (even restricting it to the cognitive dimension, e.g., Klieme & Leutner, 2006),

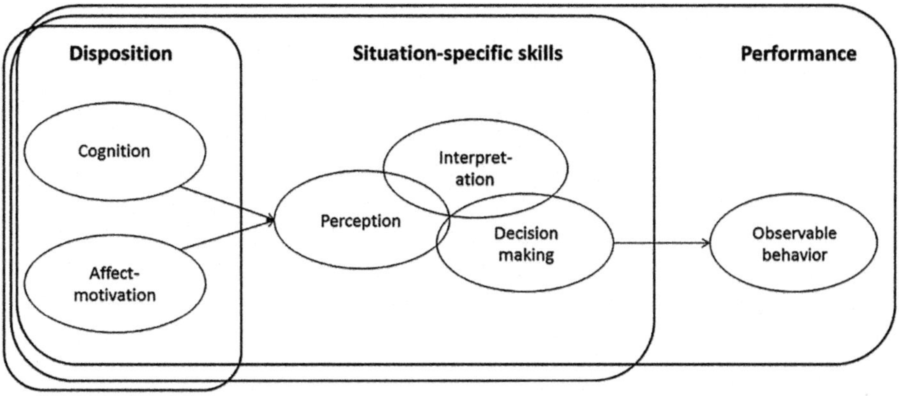

Fig. 3 Describing competence as a continuum. Reprinted with permission *from Zeitschrift für Psychologie* 2015; Vol. 223(1):3–13 © 2015 Hogrefe Publishing www.hogrefe.com, DOI:10.1027/2151-2604/a000194 (Blömeke et al., 2015)

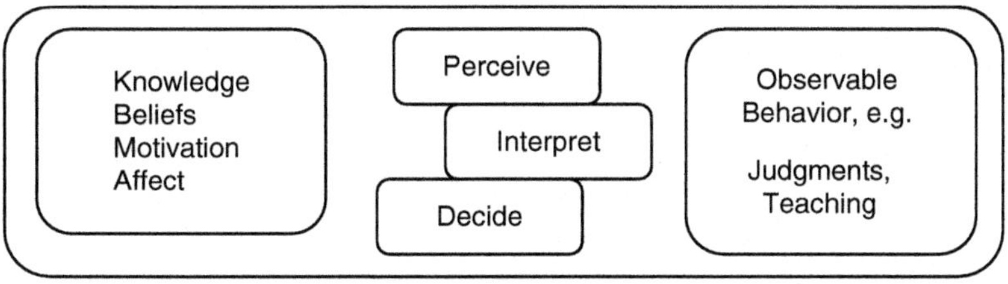

Fig. 4 Diagnostic competence (in a wider sense) as a continuum

while they consider the performance as a manifestation of this latent construct. When working in the area of competence research one should be aware of this fluctuating use of the notion of competence.

Figure 4 shows the resulting *model of diagnostic competence*, which may be considered as a framework for locating existing and future research in this area. With such a model in mind one can distinguish and classify recent approaches that investigate diagnostic competence. This will be the goal of the next sections.

1.2 Theoretical Approaches to Define the Construct of Diagnostic Competence

Although teachers' judgments on various elements of the learning process (students' learning conditions, processes and outcome, learning opportunities such as tasks, etc.) are considered relevant components of teacher activity (Ingenkamp & Lissmann, 2008; Klauer, 1978; Shavelson, 1978), no overarching theory has emerged that can be used to describe, let alone explain these judgments. Recently there have been suggestions for "working models", for example, by Aufschnaiter et al. (2015) and Herppich et al. (2017), which are still to be investigated with respect to their explanative power.

On the contrary, different research traditions have developed different approaches to describe diagnostic competence. In each of these traditions one can find all the three levels described above (diagnostic dispositions, thinking and performance), albeit with different weight and different relevance for the respective research questions. In the following sections, four influential approaches will be outlined. In a very rough categorization they may be characterized by respectively emphasizing the perspective of (1) *judgment accuracy*, (2) *competence modeling*, (3) *cognitive processes* of (social) judgment and (4) (formative) *diagnostic practices*.

(1) *Judgment accuracy*

There is a broad tradition in educational research to construe diagnostic competence as the ability to render accurate judgments on students' traits (Anders et al., 2010; Hoge & Coladarci, 1989; Schrader, 1989; Südkamp et al., 2012). Diagnostic competence is measured by quantifying the accordance of teachers' judgments with objective data (test results, student questionnaires), for example by calculating correlations or differences between solution frequencies as estimated by teachers on the one hand and as actually achieved by their students on the other hand (see Fig. 5). This method can be varied in many ways, depending on the characteristics of the data (metric or categorical, individual or group) and the focus of evaluation (average, variance or order).

This so-called "paradigm of veridicality" or "judgment accuracy paradigm" has already been suggested by Cronbach (1955), and its main findings are presented below

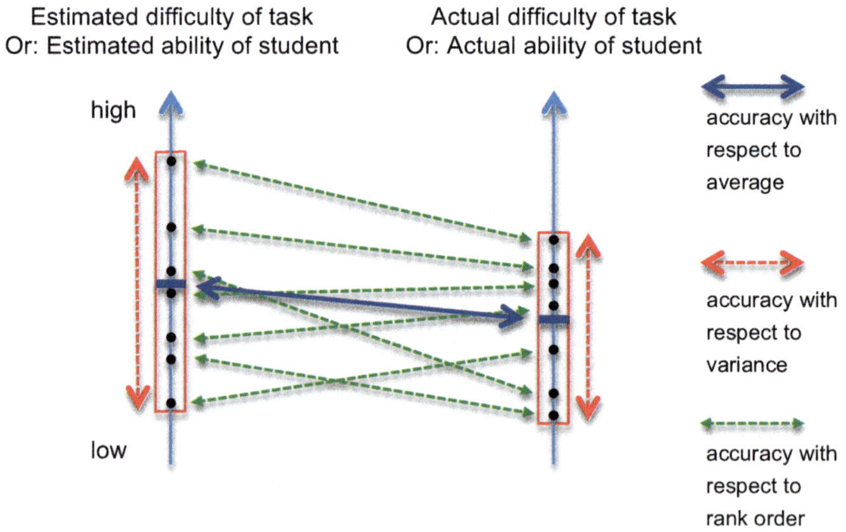

Fig. 5 Different methods of measuring judgment accuracy by comparing estimated results (judgments) with actual results. The *black dots* may either represent tasks that vary in difficulty or students who vary in ability.

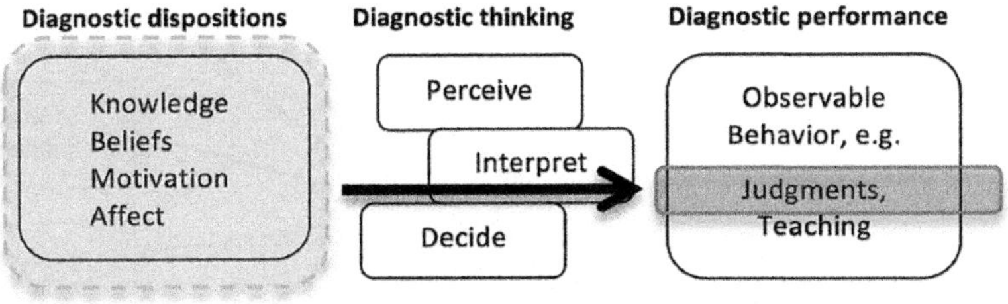

Fig. 6 Perspective of diagnostic performance. Individual dispositions and situational aspects are investigated, e.g., as factors creating a bias, resulting in a reduction in accuracy

Within the model of diagnostic competence (Fig. 4), the judgment accuracy approach focuses on the area of a specific type of diagnostic behavior, i.e., the estimation or prediction of student achievement (dark area in Fig. 6). It then strives to theoretically predict and empirically link the different measures of accuracy to conditions in the teacher dispositions (light gray area in Fig. 6) or in the situation (e.g., specific or unspecific task description).

(2) *Competence modeling*

In a broader understanding of diagnostic competence, the construct comprises the knowledge and skills of teachers, which enable them to *effectively act in all diagnostic situations* (Nitko, 2004; Schrader, 2011). This refers for example to knowledge about typical solution processes of students or about frequently occurring learning obstacles. It also comprises skills in selecting assessment procedures (formal tests, diagnostic interviews), carrying them out, evaluating them and deciding on feedback conditions – with an eye on judgment errors and biases. Such facets within a broad understanding of diagnostic competence are designated as *diagnostic expertise* (Helmke, 2010) or as *assessment literacy* (Popham, 2009). In a similar vein Aufschnaiter et al. (2015) define diagnostic competence as knowledge on diagnostic methods and on findings and theories about cognitive abilities and emotional states of learners. It is plausible that such broad conceptualizations of diagnostic competence involve the assumption of a multidimensional construct bundle (Brunner, Anders, Hachfeld, & Krauss, 2011; Lintorf et al., 2011; Praetorius, Karst, & Lipowsky, 2012; Schrader, 2013; Spinath, 2005).

This perspective on diagnostic competence as an area of teacher expertise (Bromme, 1987) naturally leads to an approach that uses approaches of "competence modeling" as proposed in the recent years (Blömeke et al., 2015; Klieme & Leutner, 2006; Herppich et al., 2017). Here, competences are defined as context-specific (cognitive) dispositions, which are functional for situations and tasks in certain domains; that is, they allow effective problem solving in these situations and domains. The selection of the domain and the specification of the situations define the area of competence in question. In this sense, diagnostic competence comprises the knowledge, beliefs and skills that enable teachers to successfully perform diagnostic activities in their domain. This perspective is reflected in the

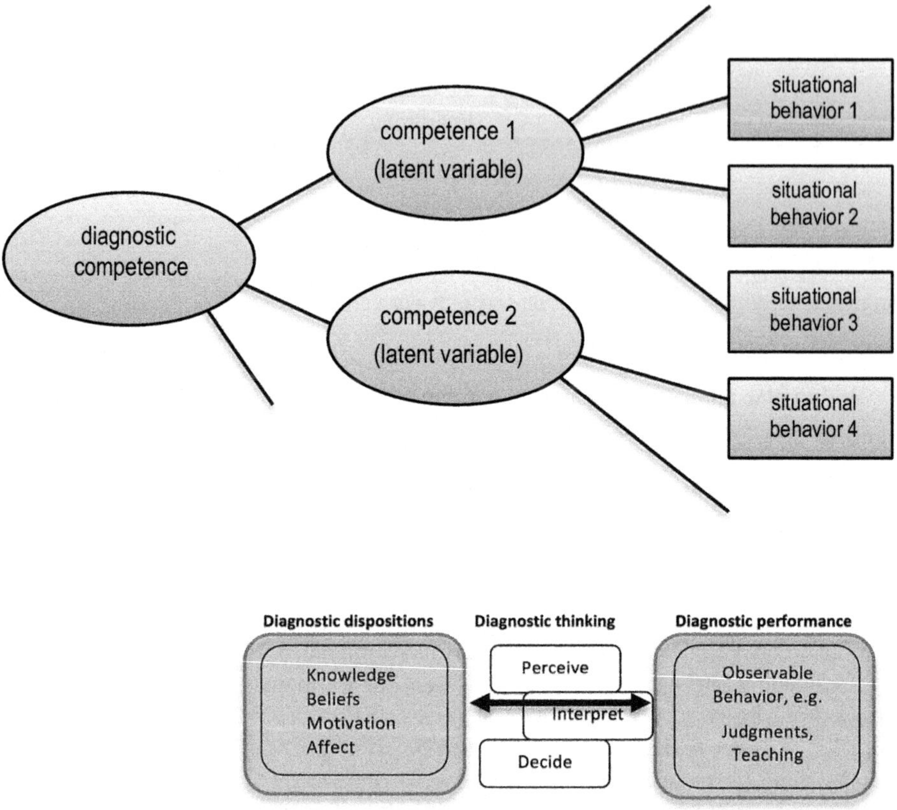

Fig. 7 Perspective of diagnostic dispositions. Diagnostic competence as a latent construct bundle

research on teacher competence – or, as preferred in the Anglo-American research tradition: knowledge and skills (e.g., Ball et al., 2008; Baumert & Kunter, 2006; Blömeke, 2011).

Diagnostic competences in this sense are latent variables that cannot be observed directly. Therefore, we need to rely on an operationalization by observable behavior and connect this behavior to these variables by a measurement procedure (e.g., an adequate psychometric model, cf. Fig. 7).

The description of situational behavior (operationalization) is not considered merely as a means for development of a measurement procedure but as constitutive for the *definition* of the competence. The definition of a competence crucially relies on the choice and operationalization of the behavior that is assumed to reflect effective diagnostic activity. This perspective emphasizes the relevance of the *context* in competence modeling. Interestingly, in the effort of Hill, Ball and Schilling (2008) to measure pedagogical content knowledge (of which diagnostic competence can be regarded as subfacet), all examples for the operationalization by test items represent situations that largely describe diagnostic activities (see Fig. 8): items 1, 2 and 3 ask

1. You are working individually with Bonny, and you ask her to count out 23 checkers, which she does successfully. You then ask her to show you how many checkers are represented by the 3 in 23, and she counts out 3 checkers. Then you ask her to show you how many checkers are represented by the 2 in 23, and she counts out 2 checkers. What problem is Bonny having here? (Mark ONE answer.)
   a. Bonny doesn't know how large 23 is.
   b. Bonny thinks that 2 and 20 are the same.
   c. Bonny doesn't understand the meaning of the places in the numeral 23.
   d. All of the above.
2. Mrs. Jackson is getting ready for the state assessment, and is planning mini-lessons for students focused on particular difficulties that they are having with adding columns of numbers. To target her instruction more effectively, she wants to work with groups of students who are making the same kind of error, so she looks at a recent quiz to see what they tend to do. She sees the following three student mistakes:

$$\begin{array}{r} {}^{1}\\ 38\\ 49\\ \underline{+65}\\ 142 \end{array} \qquad \begin{array}{r} {}^{1}\\ 45\\ 37\\ \underline{+29}\\ 101 \end{array} \qquad \begin{array}{r} {}^{1}\\ 32\\ 14\\ \underline{+19}\\ 64 \end{array}$$

(1) (II) (III)

   Which have the same kind of error? (Mark ONE answer.)
   a. I and II
   b. I and III
   c. II and III
   d. I, II, and III
3. Mr. Fitzgerald has been helping his students learn how to compare decimals. He is trying to devise an assignment that shows him whether his students know how to correctly put a series of decimals in order. Which of the following sets of numbers will best suit that purpose? (Mark ONE answer.)

a.	.5	7	.01	11.4
b.	.60	2.53	3.14	.45
c.	.6	4.25	.565	2.5

   d. Any of these would work well for this purpose. They all require the students to read and interpret decimals.
4. Consider Jill's response to a subtraction problem. How might she have gotten an answer like this?

$$\begin{array}{r} 51\\ \underline{-18}\\ 47 \end{array}$$

Fig. 8 Operationalization of diagnostic competence by items that reflect diagnostic situations (Reprinted with permission from *Journal for Research in Mathematics Education*, copyright 2008, by the National Council of Teachers of Education. All rights reserved. Hill et al., 2008, p. 400)

for the identification of frequent errors by analyzing verbal or written student data; item 4 requests to select a task with diagnostic purpose.

(3) *Cognitive processes of (social) judgment*

A different approach toward understanding diagnostic competence considers diagnostic judgments as the result of processes of social judgments, investigating the cognitive and affective influences on these processes (e.g., Krolak-Schwerdt, Böhmer, & Gräsel, 2013; or the "simulated classroom", Südkamp & Möller, 2009).

This emphasis on the cognitive processes of teachers is also reflected in approaches that explicitly model processes of perception (P), interpretation (I) and decision-making (D), such as the concepts of *noticing* (Santagata & Yeh, 2016; Sherin & van Es, 2009) and of teacher *decision-making* (Schoenfeld, 2010). These approaches are mostly used to investigate or to systematically influence teacher behavior in the midst of the instruction process.

The P-I-D-triad is chosen by Blömeke et al. (2015) to form the core categories on the skill level within the concept of competence as a continuum. These categories represent central cognitive processes that are also studied by general psychology (cf. "Thinking – judging, decision making, problem solving" by Betsch, Funke, & Plessner, 2011).

Another broad research area concerned with diagnostic judgments is situated in the context of medicine, investigating, for example, as *clinical reasoning* or *decision-making in nursing* (z. B. Croskerry, 2009). This research has little influence on the educational context so far. Ophuysen (2010) even stresses the differences between diagnostic processes in medical and educational contexts. However, the research on diagnostic practices in the medical context has a long tradition and has produced some ideas that can inspire also the educational context, in at least two regards:

1. Decision processes often bear a dual nature (Kahneman & Klein, 2009; Sloman, 1996): They rely on a so-called "system 1" that is characterized by intuitive, subconscious processes of bounded rationality that are "hard-wired" or built by experience and induction. They also may rely on a "system 2" that is characterized by analytical, deliberate processes of deductive reasoning and critical thinking that are acquired through learning. Both systems may interact in a complex way during a decision-making process; for example, a process of reaching a diagnostic judgment, whether in medicine or in education (Fig. 9). Frameworks like these may inspire the research on diagnostic competence; especially with an emphasis on cognitive processes (see Fig. 10).
2. For many years, in health science education a debate took place about the role of analytic and intuitive processes in clinical reasoning of experts: Should it be taught as a holistic ability by providing experience, or should analytical (statistical) procedures be followed (Eva, 2004)? Since diagnostic situations in the medical profession are characterized by both – scientific rigor and quick decisions in complex situations – Eva (2004) concludes that one should take into account both perspectives and therefore not only develop rigorous diagnostic procedures, but also investigate humans' heuristics and biases during diagnostic practice.

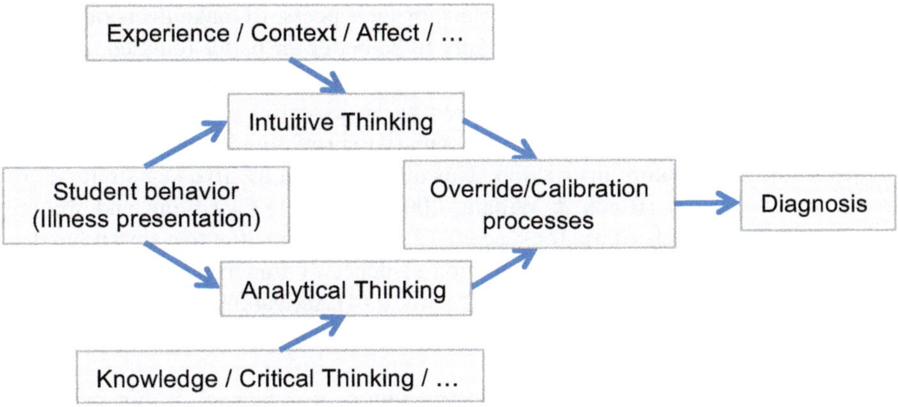

Fig. 9 A simple model for dual processes during diagnostic decision-making (for a more complex model see Fig. 12)

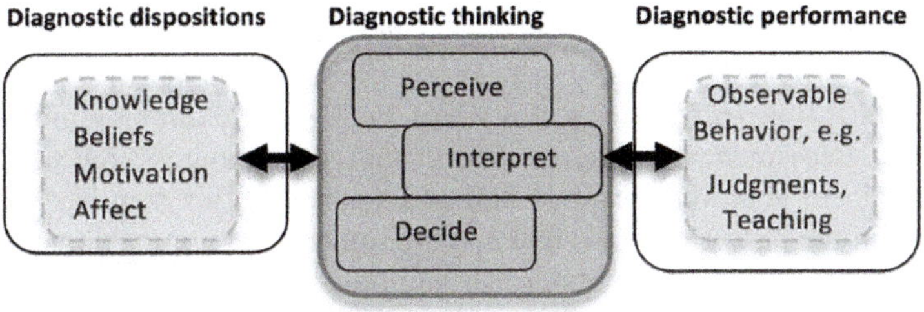

Fig. 10 Perspective of diagnostic thinking, with an emphasis on skills as perception, interpretation and decision-making

(4) *Diagnostic practices*

A rather different perspective is taken by approaches that focus on embedding diagnostic activities into classroom settings. These approaches emphasize a strong link between diagnostic activities on the one hand and ensuing instructional activities on the other hand (Abs, 2007; Aufschnaiter et al., 2015). For example, Klug, Bruder, Kelava, Spiel and Schmitz (2013) describe diagnostic competence by a multi-phase process encompassing a variety of sub-competences: planning diagnosis (pre-action phase), performing diagnosis (action phase), and implementation of the ensuing instructional measures (post-action phase).

The concept of formative assessment (Bennett, 2011; Black & Wiliam, 2003) has an even stronger focus on the integration of diagnostic and instructional activities within teaching practice (cf. also Bell's "diagnostic teaching", 1993). Black and Wiliam (2009, p. 9) define formative assessment as follows: "Practice in a classroom is formative to the extent that evidence about student achievement is elicited, interpreted, and used by teachers, learners, or their peers, to make decisions about the next steps in instruction that are likely to be better, or better founded, than the decisions they would have taken in the absence of the evidence that was elicited." Thus, diagnostic competence can be seen as an essential element of formative assessment, which is considered complementary to providing feedback and to making decisions about future instruction. This also shows in the five key strategies of formative assessment (Black & Wiliam, 2009, p. 8): (1) Clarifying and sharing learning intentions and criteria for success; (2) Engineering effective classroom discussions and other learning tasks that elicit evidence of student understanding; (3) Providing feedback that moves learners forward; (4) Activating students as instructional resources for one another, and (5) Activating students as the owners of their own learning.

Interestingly, the crucial step of "interpretation of evidence" is not explicitly included here. The focus is on eliciting and giving feedback. Further on, the authors do acknowledge that providing feedback is based on two steps (Black & Wiliam, 2009, p. 17 and p. 27): a diagnostic step (interpreting) and a prognostic step (choosing the optimum response). They go on to show that both steps are complex, but do not provide a closer look at the structure of the teachers' dispositions or skills or ways to help teachers to acquire it. Formative assessment is typically discussed from the perspective of classroom practice, that is, from a performance perspective.

Bennett (2011, p. 16ff.) takes a closer look at diagnostic thinking in the context of formative assessment. He describes the process of making inferences from evidence on student achievement as central and states that this step has often been overlooked in the past. He argues that making inferences requires domain-specific knowledge such as knowledge about learning progressions and tasks that provide evidence about those progressions in students (p. 15f.). For teachers who develop effective formative assessment practice, he concludes that they need "reasonably deep cognitive-domain understanding and knowledge of measurement fundamentals" in addition to pedagogical knowledge about the process of good teaching (p. 18f.).

Further examples of diagnostic competence as a part of complex teaching activities can be found in several chapters of this book (e.g., Biza, Nardi, & Zachariades on diagnosing and offering feedback; Koh & Chapman on designing authentic assessment tasks).

In short, in these approaches diagnostic activities are typically conceived in relation to dimensions of teaching practice, such as student-oriented teaching, individual fostering and remediation, or adaptive teaching (Beck et al., 2008; Klieme, Lipowsky, Rakoczy, & Ratzka, 2006; Pietsch, 2010). Cognitive processes or judgment accuracy are not emphasized. This is indicated in Fig. 11.

In this part of the chapter, different approaches to discuss and to investigate the construct of diagnostic competence have been described. The all can be located within the concept of *diagnostic competence as a continuum*: Each approach addresses the levels of dispositions, processes and performance in a different way and assigns to them a different role within its theoretical framework and research strategy. When addressing all the three levels and acknowledging their interconnec-

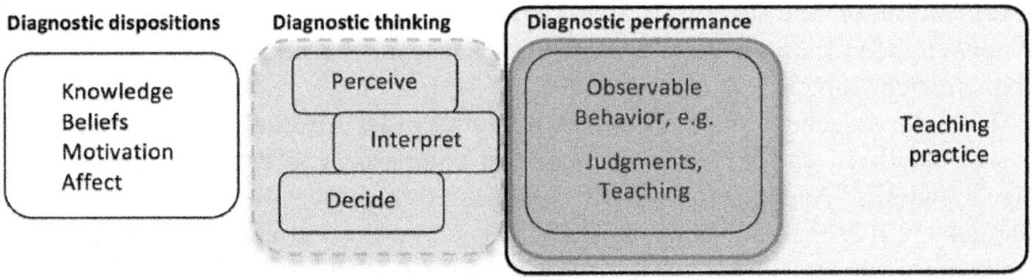

Fig. 11 The perspective of diagnostic practices. Diagnostic competence as a part of teaching practice

tions, one arrives at a more comprehensive view on diagnostic competence. This can be seen in recent research, which either follows this perspective implicitly (e.g., Hill et al., 2008; Karst, 2012) or even explicitly reflects the relevance of the three levels (e.g., Herppich et al., 2017; Heinrichs, 2015; Hoth et al., 2016). Also, each chapter in this book illustrates a specific perspective on diagnostic competence and pronounces the levels within the continuum in a specific way.

In the second part of this chapter, some of the main findings within the strands of research are collected. However, apart from guiding research, the different theoretical models described here can also inform the effort to foster diagnostic competence in pre-service and in-service teacher education. Some of those efforts can be found in several chapters in this book (e.g., Brodie, Marchant, Molefe and Chimhande on developing diagnostic competence through professional learning communities; Phelps and Spitzer on developing prospective teachers' ability to diagnose evidence of student thinking; or Stacey, Steinle, Price & Gvozdenko on building teachers' diagnostic competence and support teaching by means of an diagnostic online tool).

2 Research on Diagnostic Competence

2.1 Research from the Perspective of Diagnostic Dispositions (Knowledge, Beliefs, Motivation, Affect)

Since diagnostic competence is reflecting an important activity of teachers, it should have its place in overarching models of teacher competence. Within the influential theoretical model of Shulman (1987) diagnostic competence can be located in two categories: As part of "knowledge of learners and their characteristics" and as part of "knowledge of students' (mis)conceptions" – the former being general, the latter being a key component of pedagogical content knowledge. Ball et al. (2008) created a further differentiation by introducing the category "knowledge on content and students" (KCS) which might be equated with diagnostic competence (for a differentiated argument see below). It comprises the anticipation of students' thinking and motivation and of their specific difficulties as well as the interpretation of "the students' emerging and incomplete thinking" which requires "knowledge of common student conceptions and misconceptions about particular mathematical content" (Ball et al., 2008, p. 401).

To capture teacher competences, often a set of tasks is used that represents diagnostic situations, and which is evaluated in comparison to expert solutions (see Fig. 8 and e.g., Anders et al., 2010; Brunner et al., 2011; Döhrmann, Kaiser, & Blömeke, 2012; Hill, Schilling, & Ball, 2004). More recent approaches use video vignettes to capture teacher competences in a context that is closer to the classroom situation (Kaiser et al., 2017; Lindmeier, 2011). Consequently, with the ongoing research on teacher competences and especially on pedagogical content knowledge (Depaepe et al., 2013) also some indicators for diagnostic competence emerge by virtue of the description of teachers' successful acting in diagnostic situations (see the example in Fig. 8). Therefore, models of diagnostic competence should be distinguished not by their measurement strategies, but in terms of the narrowness or uniqueness of the construct within the respective theoretical framework. In this sense, diagnostic competence within the models presented in the following plays different roles:

(a) It is understood as a subfacet of other constructs (pedagogical knowledge (*PK*) and pedagogical content knowledge (*PCK*)).
(b) It is regarded as a unique dimension of teachers' competencies or knowledge.
(c) It is not specified distinctly but represented implicitly by certain aspects of teacher acting or thinking.

These different viewpoints and the state of research will be outlined separately in the following.

(a) *Diagnostic competence as a subfacet of PK and PCK*

Diagnostic competence is regarded as a significant component of teaching but not as a unique dimension and not separately modeled in studies dealing with professional knowledge of teachers (e.g., COACTIV: Kunter, Baumert, & Blum, 2011; TEDS-M: Döhrmann et al., 2012). To assess diagnostic competence the studies comprise items on the knowledge of students' thinking, task properties or ability testing (e.g., Brunner et al., 2011; Hoth et al., 2016). Thus, diagnostic competence is seen as a collection of certain aspects of teachers' acting that contribute to pedagogical knowledge and pedagogical content knowledge.

In-depth explorations by qualitative studies gain insights into diagnostic thinking of teachers while solving items on PK and PCK: Teachers may focus on task content or student thinking processes (Philipp & Leuders, 2014; Philipp, chapter "Diagnostic Competences of Mathematics Teachers with a View to Processes and Knowledge Resources" in this book; Reinhold, chapter "Revealing and Promoting Pre-Service Teachers' Diagnostic Strategies in Mathematical Interviews with First-Graders" in this book) or on errors in students' solutions (Heinrichs, chapter "Diagnostic Competence for Dealing with Students' Errors: Fostering Diagnostic Competence in Error Situations" in this book). Teachers with a focus on task content tend to provide ratings for the students' achievement while other teachers with a stronger focus on students' perspectives or thinking processes provide more arguments for fostering students (Hoth et al., 2016). Such analyses reveal a complexity of teacher behavior that can be used for better understanding, modeling and investigating cognitive processes connected to diagnostic competence (see Sect. 2.3).

(b) *Knowledge about (domain specific) student thinking as a unique dimension of diagnostic competence*

Shulman (1986) conceives pedagogical content knowledge (PCK) as an amalgam of content knowledge (CK) and pedagogical knowledge (PK). He highlights among other aspects of PCK the "understanding of what makes the learning of specific topics easy or difficult: the conceptions and preconceptions that students of different ages and background bring with them to the learning of those most frequently taught topics and lessons" (Shulman, 1986, p. 9). Inspired by this, Hill et al. (2008) further elaborate the terminology of PCK. They clearly separate it from subject matter knowledge. Subject matter knowledge instead comprises specialized content knowledge (SCK) on the one hand and common content knowledge (CCK) and knowledge on the mathematical horizon on the other hand. PCK is considered to have at least three dimensions: knowledge of content and students (KCS), knowledge of content and teaching (KCT) and knowledge of curriculum. KCS is used in situations that involve recognizing the affordances of specific content and the learning conditions and processes of particular learners, whereas KCT includes specific teaching strategies (p. 375). Hill et al. (2008, p. 380) further specify the KCS dimension by the following settings:

(I) Common student errors: Teachers should be able to identify and explain common errors having an idea for what errors arise with what content.
(II) Students' understanding of content: Teachers should analyze students' learning outcomes as sufficient or not sufficient to indicate understanding and differentiate between different levels of understanding.
(III) Student developmental sequence: Teachers also ought to be aware of problem types, topics or mathematical activities that students struggle with, depending on age.
(IV) Common student computational strategies: Finally, experienced teachers will be familiar with typical strategies of their students to solve mathematical problems.

Hill et al. (2008) can show by constructing and evaluating appropriate tests that KCS is not only theoretically but also empirically separable from CK and can be fostered by trainings focusing on students' understanding. A deeper understanding of teacher diagnostic reasoning is gained by cognitive interviews during task solving: it is obvious that teachers face heuristic strategies based on (implicit) knowledge and/or analytical strategies for judging tasks or students. Earlier, Magnusson, Karjcik and Borko (1999) suggested a similar concept for science teaching: *knowledge of students' understanding in science*. However, empirical evidence for their construct is only available in single case studies and still needs to be replicated (Park & Oliver, 2008).

(c) *Assessing aspects of teacher acting or thinking*

Although there are hardly any attempts to comprehensively conceptualize and assess diagnostic competence, one can find many studies that concentrate on single aspects of diagnostic competence by referencing to specific situations. For example, Karst (2012) developed a model that uses the alignment of learners' abilities and mathematical tasks as a central idea. Within this framework they distinguish between diagnostic situations in which the teacher either estimates: (1) the average difficulty of tasks, (2) the average achievement of students or (3) the ability of specific students to solve specific tasks. The results show only low correlations between these situations (measured within the accuracy approach, see above), so that the competences in these situations may be assumed to be distinguishable.

Heinrichs and Kaiser (Heinrichs, 2015 and Heinrichs and Kaiser, chapter "Diagnostic Competence for Dealing with Students' Errors – Fostering Diagnostic Competence in Error Situations" in this book) investigate the diagnostic competence in so-called "error situations", in which teachers are expected to (1) perceive a students' error, (2) hypothesize about the causes of the error and (3) decide on how to deal with the error. These ideal steps of the diagnostic process correspond to cognitive processes of perception, interpretation and decision-making and therefore already bear aspects of approaches that consider processes of diagnostic thinking (see next section). The errors are presented by means of video vignettes with respect to several areas of the mathematics curriculum. The authors show that these competences can be fostered in teacher education, and that they correlate with instructivist vs. constructivist teacher beliefs.

A much narrower conceptualization of diagnostic competence is introduced by Krauss and Brunner (2011). They measure how fast and how correct teachers detect mathematical errors in students' solutions and thus try to measure a "quick-assessment-competence". Among all factors considered, this competence can best be explained empirically by teachers' subject matter knowledge with respect to school-level mathematics. However, a study that links this operationalization with other measurements of diagnostic competence within the sample (Brunner et al., 2011) is still missing (for more recent results see Binder, Krauss, Hilbert, Brunner, Anders & Kunter, chapter "Diagnostic Skills of Mathematics Teachers in the COACTIV Study" in this book).

Leuders and Leuders (2013) distinguish between specific and non-specific judgments on student solutions and find that teachers provide richer and more specific judgments when they solve the tasks by themselves before judging the solutions.

Further conceptualization of diagnostic competence via the specification of diagnostic situations can be found in several chapters of this book (e.g., Clarke, Roche & Clarke, chapter "Supporting Mathematics Teachers' Diagnostic Competence Through the Use of One-to-One, Task-Based Assessment Interviews" on conducting task-based assessment interviews; Hiebert, Morris & Spitzer, chapter "Diagnosing Learning Goals: An Often-Overlooked Teaching Competency" on diagnosing learning goals).

Concluding this section, one may say that many studies assess aspects or facets of diagnostic competence by including indicators beside and beyond assessment accuracy to operationalize a more profound pattern of judgment processes and products (e.g., Ball et al., 2008; Busch, Barzel, & Leuders, 2015; Heinrichs, 2015; Streit, Rüede, & Weber, 2015; see also the chapters of this book). However, no coherent picture is emerging for two reasons. Studies cannot be interpreted within a common theoretical framework and therefore there is no opportunity yet to develop a 'nomological net' (Cronbach & Meehl, 1955). Furthermore the identification of influencing factors by regression-like analyses only reveals fragmentary glimpses on cognitive processes that underlie diagnostic judgments. Therefore a more systematic and a priori consideration of models for diagnostic thinking processes should be envisioned to improve the situation.

2.2 Research from the Perspective of Diagnostic Thinking (Diagnostic Skills)

Teacher behavior is the product of many individual, situational and contextual factors during decision-making (Schoenfeld, 2011). Findings from the research outlined above yield correlations of diagnostic competences with various factors but they often lack explanations for the development and the underlying reasons of these correlational patterns. Little is known about the decision-making process itself. Decision-making in educational contexts can be located within the framework of psychological decision-making research (Hastie & Dawes, 2001, especially *social cognition*; Bless, Fiedler, & Strack, 2004). This research, for example, reveals certain kinds of judgment biases connected to classroom assessment, which are well known in educational settings (Ingenkamp & Lissmann, 2008). Drawing from that, we focus on theories that might help to explain decision-making rather than describing it.

Brunswik's *lens model of perception* (1955) can contribute to describe the teachers' decision-making process. According to Brunswik, students' characteristics are *distal* traits that cannot be observed directly. Instead, the properties can be revealed through observations of more or less *proximal* indicators of behavior. The decision-making process of the teacher can thus be separated into two components: First, the perception and professional vision of behavioral indicators (Brunswik calls them *cues*) and second, the valid interpretation of those cues with respect to the own expertise. Both components are essential for decision-making, resulting in a huge diversity in diagnostic competences in teachers: They notice different information and interpret them differently from their respective knowledge background. Thus, relevant or irrelevant cues are used to reason correctly or incorrectly, which leads to diverse decisions about student's competences and characteristics. Decision-making might additionally be biased by world knowledge, implicit theories or beliefs, etc. Even affective factors can determine decision- making (Blömeke et al., 2015); mood- or stress-related aspects would bias perception and reasoning in decision-making processes (Forgas, 2000). Although Brunwik's lens model offers a theoretical approach for the investigation of judgment processes and although it has been widely applied in general psychology (cf. Karelaia & Hogarth, 2008), it has only rarely been used in the area of education (e.g., Rothenbusch, 2017).

In general, cognitive psychology comprises two distinctive processes of decision-making that run partly parallel: Systematic processing and heuristic processing. Systematic processing involves a comprehensive and systematic search, or the deliberate generation of judgment-relevant information. Heuristic processing means a decision in a narrower sense, based on the teachers' implicit knowledge and heuristics (Hoffrage, Hertwig, & Gigerenzer, 2005). During systematic information processing the relevant cues are investigated in an analytical manner searching for decision-relevant cues. However, there are individual variations in how many cues are considered, when the search for information is abandoned and which way the identified cues are weighted in the decision-making process (Swets, Daws, & Monahan, 2000). This systematic process consumes more time and capacity than heuristic processing, which is based on inherent heuristic strategies or learnt knowledge. Heuristics is valid if relevant information is stored in memory, which can be retrieved in the diagnostic situation (Chen, Duckworth, & Chaiken, 1999). Within this heuristic processing a person might target cues that are helpful for decision-making (matching heuristics). Once a "critical" cue is found the decision-making process will be terminated and additional or more significant cues will be ignored, which can ultimately lead to biased decisions (Dhami & Harries, 2001).

Prominent dual-process models include both perspectives (Kahneman, 2003; heuristic-systematic model, Chen & Chaiken, 1999) to describe social decision-making processes and add further components such as motivation and responsibility of the person making the decision (Chen et al., 1999; Lerner & Tetlock, 1999). It can be assumed that the time-consuming and resource-demanding analytical strategies will only be used when highly motivated persons encounter inconsistent information that does not meet their individual heuristics. For example, a decision on having the most suitable school track for a child could be a possible scenario. Heuristic strategies instead will be preferred when the diagnosis situation provides familiar information that is in line with own heuristics or when the person is less motivated.

To understand diagnostic judgments, not only in the educational context, there is a need to use and develop more elaborated decision-making models, especially in medicine or medical assessment (Chapman & Sonnenberg, 2000). In this area heuristic-intuitive and systematic-analytical models were investigated and integrated into dual-process models, such as the model of Croskerry (2009). In his model, the relevant cues (in a teaching setting it would be, e.g., task characteristics and properties of the students) would be more or less observed and processed depending on overlearning and practice; during recurrent diagnostic situations the internal processor is getting calibrated. It can be assumed that findings from the medical field can be useful for educational assessment. In these, the teacher also has to reason about more or less obvious cues, weighing cues based on his/her expertise and practice, and finally decide (Fig. 12).

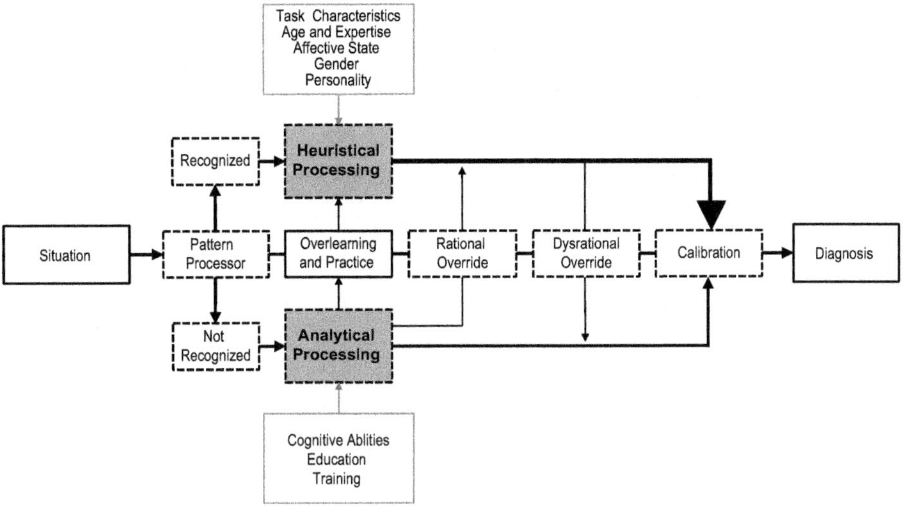

Fig. 12 Model for diagnostic reasoning based on pattern recognition and dual-process theory. Simplified version of Croskerry (2009, p. 1024)

Croskerry's model is linear, running from left to right. The initial presentation of the diagnostic situation has to be recognized by the observer. If it is recognized, the parallel, fast, automatic processes engage (heuristic); if it is not recognized, the slower, analytical processes engage instead. Determinants of heuristic and analytical processes are shown in dotted-line boxes. Repetitive processing in the analytical system leads to recognition and default situations to heuristic processing system. Either system may override the other. Both system outputs pass into a calibrator in which interaction may or may not occur to produce the final diagnosis (Croskerry, 2009, p. 1024).

There are only very few efforts that explicitly try to elucidate the cognitive processes activated during diagnostic judgments within cognitive frameworks. Nickerson (1999) presents a general model based on a huge body of research on people's understanding of the knowledge of others and on processes of imputing other people's knowledge. Experts consistently seem to use their own knowledge as a default model. By using information about a "random" other they adapt their model and further adapt continuously when they gain information on "specific" others. In Nickerson's model own knowledge is set to be an anchor that would be modified through content knowledge about other persons in teaching situations and the experiences the teacher makes in teaching settings. A teacher will be more or less aware of his own knowledge with all its characteristics and unusual aspects. Together with a heuristic knowledge about a random other's knowledge and general knowledge a specific model of other's knowledge (in this case the knowledge of specific students) is constructed and would be updated by on-going information. The resulting working model can be used to diagnose students' abilities and challenges (Fig. 13).

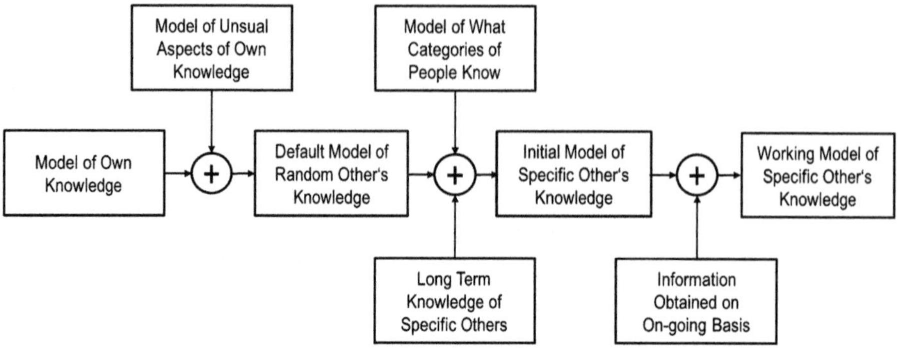

Fig. 13 Illustrating the base of one's working model of specific other's knowledge (Nickerson, 1999, p. 740)

This heuristic of anchoring and adjustment (Tversky & Kahneman, 1974) explains the ubiquitous phenomenon of overestimation, called "curse of expertise" (e.g., Camerer, Loewenstein, & Weber, 1989). This model has only recently been used to explain diagnostic processes in pedagogical contexts (Ostermann, Leuders, & Nückles, 2015) and must be adapted carefully since underestimation tendencies have been observed as well (see also the chapter by Ostermann, "Factors Influencing the Accuracy of Diagnostic Judgments" in this book).

The approach helps to interpret diverse bias in professional teacher perception such as the expert blind spot (Nathan & Koedinger, 2000; Nathan & Petrosino, 2003). Until now the approach is not being used in research on teachers' diagnostic competences, although Morris, Hiebert and Spitzer (2009) showed in accordance with the model, how decompression of knowledge (in the topic "analyzing minor goals of a task") could foster diagnostic judgments (see also the chapter by Hiebert, Morris & Spitzer, "Diagnosing Learning Goals: An Often Overlooked Teaching Competency" and the chapter of Philipp, "Diagnostic Competences of Mathematics Teachers with a View to Processes and Knowledge Resources" in this book).

In summary, it is evident that a special focus in this line of research is on (cognitive and affective) teacher characteristics, aiming to explain differences in teacher outcome. Although available studies indicate correlations between teacher and student characteristics the mechanism of decision-making is still a mostly neglected area of research. The models mentioned above could help to reveal the underlying decision-making processes from a theoretical perspective, supporting studies using, for example, experiments, cognitive interviews or eye tracking.

2.3 Research from the Perspective of Diagnostic Performance

Diagnostic competence is often seen as related to student achievement and for that purpose defined as the ability of teachers to accurately assess students' performance (Schrader, 2006). The accuracy is measured in many ways (Schrader & Helmke,

1987), which can be ranked according to their specificity (cf. Südkamp et al., 2012): judging students by a rating scale (lowest specificity), classifying students into a rank order, predicting grades of a standardized achievement test, predicting the number of correct responses and indicating expected responses (highest specificity). Most often the correspondence between the teachers' judgment and the students' actual achievement is reported by means of a correlation coefficient. The research summarized in Hoge and Coladarci (1989, 16 studies) and Südkamp et al. (2012, 75 studies) showed an averaged correspondence of about 0.66 resp. 0.63, indicating that teachers do quite well in judging their students' achievement, but there is considerable variation among the teachers (typically 0.3–0.9). The diverse coefficients used in this strand of research can merely be regarded as indicators for a latent construct, which is rather poorly defined in terms of cognitive processes (Schrader, 2011).

Findings from previous research show that students' characteristics (e.g., achievement level), judgment characteristics (e.g., specific vs. global) or test formats (e.g., formative vs. summative assessments) could not reveal a coherent pattern over domains that help to explain diagnostic processing (Südkamp, Kaiser, & Möller, 2012). So, the question which characteristics best explain competence differences between teachers has not been answered to date. Properties of teachers such as cognitive abilities (Kaiser, Retelsdorf, Südkamp, & Möller, 2012) or beliefs (Shavelson & Stern, 1981) are regarded to be moderator variables with low impact. The ability level (Carr & Kurtz-Costes, 1994) and additional cognitive factors (Schrader & Helmke, 1999) are student characteristics that are discussed as possible moderators of diagnostic competence. Moreover, personality similarity between teachers and their students is a predictor of diagnostic competence (Rausch, Karing, Dörfler, & Artelt, 2015) from a social psychological point of view. However, the predictive power of the abovementioned variables is still low.

Most studies on judgment accuracy use correlations between students' achievement and teachers' assessments as indicators for diagnostic competences. To improve the situation, research that explicitly draws on models for the mechanisms that underlie the judgment process and that may thus explain differences in judgment accuracy is needed. Due to the large variety of conceptualizations, meta-studies (like, e.g., Südkamp et al., 2012) cannot yield substantial findings on the influence of certain variables, such as the teacher knowledge. Instead, a systematic experimental variation (such as instruction on student errors as in Ostermann, Leuders, & Nückles, 2017) may be a path for further research on diagnostic competences via judgment accuracy.

References

Abs, H. J. (2007). Überlegungen zur Modellierung diagnostischer Kompetenz bei Lehrerinnen und Lehrern. In M. Lüders & J. Wissinger (Eds.), *Forschung zur Lehrerbildung* (pp. 63–84). Münster, Germany: Waxmann.

Anders, Y., Kunter, M., Brunner, M., Krauss, S., & Baumert, J. (2010). Diagnostische Fähigkeiten von Mathematiklehrkräften und die Leistungen ihrer Schülerinnen und Schüler. *Psychologie in Erziehung und Unterricht*, *3*, 175–193.

Artelt, C., & Gräsel, C. (2009). Diagnostische Kompetenz von Lehrkräften. *Zeitschrift für Pädagogische Psychologie, 23*(3–4), 157–160.
von Aufschnaiter, C., Cappell, J., Dübbelde, G., Ennemoser, M., Mayer, J., Stiensmeier-Pelster, J., … Wolgast, A. (2015). Diagnostische Kompetenz. Theoretische Überlegungen zu einem zentralen Konstrukt der Lehrerbildung. *Zeitschrift für Pädagogik, 61*, 738–758.
Ball, D. L., Thames, M. H., & Phelps, G. (2008). Content knowledge for teaching: What makes it special? *Journal of Teacher Education, 59*(5), 389–407.
Baumert, J., & Kunter, M. (2006). Stichwort: Professionelle Kompetenz von Lehrkräften. *Zeitschrift für Erziehungswissenschaft, 9*, 469–520.
Baumert, J., Kunter, M., Blum, W., Brunner, M., Voss, T., Jordan, A., … Tsai, Y.-M. (2010). Teachers' mathematical knowledge, cognitive activation in the classroom, and student progress. *American Educational Research Journal, 47*(1), 133–180.
Beck, E., Baer, M., Guldimann, T., Bischoff, S., Brühwiler, C., Müller, P., … Vogt, F. (2008). *Adaptive Lehrkompetenz. Analyse und Struktur, Veränderung und Wirkung handlungssteuernden Lehrerwissens*. Münster, Germany: Waxmann.
Bell, A. (1993). Some experiments in diagnostic teaching. *Educational Studies in Mathematics, 24*(1), 115–137.
Bennett, R. E. (2011). Formative assessment: A critical review. *Assessment in Education: Principles, Policy & Practice, 18*(1), 5–25.
Berliner, D. (1994). Expertise: The wonder of exemplary performances. In J. Mangieri & C. Block (Eds.), *Creating powerful thinking in teachers and students* (pp. 161–186). Ft. Worth, TX: Holt, Rinehart and Winston.
Betsch, T., Funke, J., & Plessner, H. (2011). *Denken - Urteilen, Entscheiden, Problemlösen: Mit 14 Tabellen*. Berlin, Germany: Springer.
Black, P., & Wiliam, D. (2003). 'in praise of educational research': Formative assessment. *British Educational Research Journal, 29*, 623–637.
Black, P., & William, D. (2009). Developing the theory of formative assessment. *Educational Assessment, Evaluation and Accountability, 21*, 5–31.
Bless, H., Fiedler, K., & Strack, F. (2004). *Social cognition. How individuals construct social reality*. Hove, UK: Psychology Press.
Blömeke, S. (2011). Forschung zur Lehrerbildung im internationalen Vergleich. In E. Terhart, H. Bennewitz, & M. Rothland (Eds.), *Handbuch der Forschung zum Lehrerberuf* (pp. 345–361). Münster, Germany: Waxmann.
Blömeke, S., Gustafsson, J.-E., & Shavelson, R. (2015). Beyond dichotomies: Competence viewed as a continuum. *Zeitschrift für Psychologie, 223*, 3–13.
Bromme, R. (1987). Der Lehrer als Experte - Entwurf eines Forschungsansatzes. Denken und Wissen von Lehrern aus der Perspektive neuerer Forschung zum Problemlösen. In H. Neber (Ed.), *Angewandte Problemlösepsychologie* (pp. 127–151). Münster, Germany: Aschendorff.
Bromme, R. (1997). Kompetenzen, Funktionen und unterrichtliches Handeln des Lehrers. In F. E. Weinert (Ed.), *Psychologie des Unterrichts und der Schule* (pp. 177–212). Göttingen u.a: Hogrefe (Enzyklopädie der Psychologie; D, 1, 3).
Brunner, M., Anders, Y., Hachfeld, A., & Krauss, S. (2011). Diagnostische Fähigkeiten von Mathematiklehrkräften. In M. Kunter, J. Baumert, W. Blum, U. Klusmann, S. Krauss, & M. Neubrand (Eds.), *Professionelle Kompetenz von Lehrkräften – Ergebnisse des Forschungsprogramms COACTIV* (pp. 215–234). Münster, Germany: Waxmann.
Brunswik, E. (1955). Representative design and probabilistic theory in a functional psychology. *Psychological Review, 62*, 193–217.
Busch, J., Barzel, B., & Leuders, T. (2015). Die Entwicklung eines kategorialen Kompetenzmodells zur Erfassung diagnostischer Kompetenzen von Lehrkräften im Bereich Funktionen. *Journal für Mathematik-Didaktik, 36*(2), 315–338.
Camerer, C., Loewenstein, G., & Weber, M. (1989). The curse of knowledge in economic settings: An experimental analysis. *Journal of Political Economy, 97*(5), 1232–1254.
Carpenter, T. P., Fennema, E., Peterson, P. L., Chiang, C.-P., & Loef, M. (1989). Using knowledge of children's mathematics thinking in classroom teaching: An experimental study. *American Educational Research Journal, 26*, 499–531.

Carr, M., & Kurtz-Costes, B. E. (1994). Is being smart everything? The influence of student achievement on teachers' perceptions. *British Journal of Educational Psychology, 64*, 263–276.

Chapman, G. B., & Sonnenberg, F. A. (2000). *Decision making in health care: Theories, psychology and applications. Cambridge series on judgment and decision making*. Cambridge, UK: Cambridge University Press.

Chen, S., & Chaiken, S. (1999). The heuristic-systematic model in its broader context. In S. Chaiken & Y. Trope (Eds.), *Dual process theories in social psychology* (pp. 73–96). New York, NY: Guilford.

Chen, S., Duckworth, K., & Chaiken, S. (1999). Motivated heuristic and systematic processing. *Psychological Inquiry, 10*(1), 44.

Cronbach, L. J. (1955). Processes affecting scores on "understanding of others" and "assumed similarity". *Psychological Bulletin, 52*, 177–193.

Cronbach, L. J., & Meehl, P. E. (1955). Construct validity in psychological tests. *Psychological Bulletin, 52*, 281–303.

Croskerry, P. (2009). A universal model of diagnostic reasoning. *Academic Medicine, 84*(8), 1022–1028.

Darling-Hammond, L. (2000). Teacher quality and student achievement: A review of state policy evidence. *Education Policy Analysis Archives, 8*(1), 1–48.

Demaray, M. K., & Elliott, S. N. (1998). Teachers' judgments of students' academic functioning: A comparison of actual and predicted performances. *School Psychology Quarterly, 13*(1), 8–24.

Depaepe, F., Verschaffel, L., & Kelchtermans, G. (2013). Pedagogical content knowledge: A systematic review of the way in which the concept has pervaded mathematics educational research. *Teaching and Teacher Education, 34*, 12–25.

Dhami, M., & Harries, C. (2001). Fast and frugal versus regression models of human judgment. *Thinking and Reasoning, 7*, 5–27.

Döhrmann, M., Kaiser, G., & Blömeke, S. (2012). The conceptualisation of mathematics competencies in the international teacher education study TEDS-M. *The International Journal on Mathematics Education, 44*(3), 325–340.

Eva, K. (2004). What every teacher needs to know about clinical reasoning. *Medical Education, 39*, 98–106.

Forgas, J. P. (2000). Affect and information processing strategies: An interactive relationship. In J. P. Forgas (Ed.), *Feeling and thinking: The role of affect in social cognition* (pp. 253–282). New York, NY: Cambridge University Press.

Hastie, R., & Dawes, R. M. (2001). *Rational choice in an uncertain world: The psychology of judgment and decision making*. Thousand Oaks, CA: Sage Publications.

Heinrichs, H. (2015). *Diagnostische Kompetenz von Mathematik-Lehramtsstudierenden: Messung und Förderung*. Heidelberg, Germany: Springer.

Helmke, A. (2010). *Unterrichtsqualität und Lehrerprofessionalität. Diagnose, Evaluation und Verbesserung des Unterrichts*. Seelze, Germany: Klett-Kallmeyer.

Hengartner, E., Hirt, U., & Wälti, B. (2007). *Lernumgebungen für Rechenschwache bis Hochbegabte: Natürliche Differenzierung im Mathematikunterricht*. Zug, Switzerland: Klett und Balmer.

Herppich, S., Praetorius, A.-K., Hetmanek, A., Glogger, I., Ufer, S., Leutner, D., … Südkamp, A. (2017). Ein Arbeitsmodell für die empirische Erforschung der diagnostischen Kompetenz von Lehrkräften. In A. Südkamp & A.-K. Praetorius (Eds.), *Diagnostische Kompetenz von Lehrkräften. Theoretische und methodische Weiterentwicklungen*. Münster, Germany: Waxmann.

Hill, H. C., Ball, D. L., & Schilling, S. G. (2008). Unpacking pedagogical content knowledge: Conceptualizing and measuring teachers' topic-specific knowledge of students. *Journal for Research in Mathematics Education, 39*(4), 372–400.

Hill, H. C., Schilling, S. G., & Ball, D. L. (2004). Developing measures of teachers' mathematics knowledge for teaching. *Elementary School Journal, 105*, 11–30.

Hodgen, J. (2011). Knowing and identity: A situated theory of mathematics knowledge in teaching. In T. Rowland & K. Ruthven (Eds.), *Mathematical knowledge in teaching* (pp. 27–42). Dordrecht: Springer Netherlands.

Hoffrage, U., Hertwig, R., & Gigerenzer, G. (2005). Die ökologische Rationalität einfacher Entscheidungs- und Urteilsheuristiken. In H. Siegenthaler (Ed.), *Rationalität im Prozess kultureller Evolution* (pp. 65–89). Tübingen: Mohr Siebeck.

Hoge, R. D., & Coladarci, T. (1989). Teacher-based judgments of academic achievement: A review of literature. *Review of Educational Research, 59*(3), 297–313.

Hoth, J., Döhrmann, M., Kaiser, G., Busse, A., König, J., & Blömeke, S. (2016). Diagnostic competence of primary school mathematics teachers during classroom situations. *The International Journal on Mathematics Education, 48*(1–2), 41–54.

Ingenkamp, K., & Lissmann, U. (2008). *Lehrbuch der pädagogischen Diagnostik* (6th ed.). Weinheim, Germany/Basel, Switzerland: Beltz.

Kahneman, D. (2003). A perspective on judgment and choice. *American Psychologist, 58*(9), 697–720.

Kahneman, D., & Klein, G. (2009). Conditions for intuitive expertise. *American Psychologist, 64*, 515–526.

Kaiser, G., Blömeke, S., König, J., Busse, A., Döhrmann, M., & Hoth, J. (2017). Professional competencies of (prospective) mathematics teachers – Cognitive versus situated approaches. *Educational Studies in Mathematics, 94*(2), 161–182.

Kaiser, J., Retelsdorf, J., Südkamp, A., & Möller, J. (2013). Achievement and engagement: How student characteristics influence teacher judgments. *Learning and Instruction, 28*(0), 73–84.

Karelaia, N., & Hogarth, R. M. (2008). Determinants of linear judgment: A meta-analysis of lens model studies. *Psychological Bulletin, 134*, 404–426.

Karst, K. (2012). *Kompetenzmodellierung des diagnostischen Urteils von Grundschullehrern*. Münster, Germany: Waxmann.

Klauer, K. J. (1978). Perspektiven pädagogischer Diagnostik. In K. J. Klauer (Ed.), *Handbuch der Pädagogischen Diagnostik* (pp. 3–14). Düsseldorf, Germany: Schwann.

Klieme, E., & Leutner, D. (2006). Kompetenzmodelle zur Erfassung individueller Lernergebnisse und zur Bilanzierung von Bildungsprozessen. Beschreibung eines neu eingerichteten Schwerpunktprogramms der DFG. *Zeitschrift für Pädagogik, 52*(6), 876–903.

Klieme, E., Lipowsky, F., Rakoczy, K., & Ratzka, N. (2006). Qualitätsdimensionen und Wirksamkeit von Mathematikunterricht. Theoretische Grundlagen und ausgewählte Ergebnisse des Projekts 'Pythagoras'. In M. Prenzel & L. Aloi-Näcke (Eds.), *Untersuchungen zur Bildungsqualität von Schule. Abschlussbericht des DFG-Schwerpunktprogramms* (pp. 127–146). Münster, Germany: Waxmann.

Klug, J., Bruder, S., Kelava, A., Spiel, C., & Schmitz, B. (2013). Diagnostic competence of teachers: A process model that accounts for diagnosing learning behavior tested by means of a case scenario. *Teaching and Teacher Education, 30*, 38–46.

Koeppen, K., Hartig, J., Klieme, E., & Leutner, D. (2008). Current issues in competence modeling and assessment. *Zeitschrift für Psychologie, 216*, 61–73.

Krauss, S., & Brunner, M. (2011). Schnelles Beurteilen von Schülerantworten: Ein Reaktionszeittest für Mathematiklehrer/innen. *Journal für Mathematikdidaktik, 32*, 233–251.

Krolak-Schwerdt, S., Böhmer, M., & Gräsel, C. (2013). The impact of accountability on teachers' assessments of student performance: A social cognitive analysis. *Social Psychology of Education, 16*, 215–239.

Kultusministerkonferenz. (2014). *Standards Lehrerbildung KMK-Arbeitsgruppe*. Available at www.kmk.org/fileadmin/veroeffentlichungen_beschluesse/2004/2004_12_16-Standards-Lehrerbildung.pdf. 1 June 2017.

Kunter, M., Baumert, J., & Blum, W. (Eds.). (2011). *Professionelle Kompetenz von Lehrkräften: Ergebnisse des Forschungsprogramms COACTIV*. Münster, Germany: Waxmann.

Lerner, J. S., & Tetlock, P. E. (1999). Accounting for the effects of accountability. *Psychological Bulletin, 125*(2), 255–275.

Leuders, J., & Leuders, T. (2013). Improving diagnostic judgement of preservice teachers by reflective task solution. In A. M. Lindmeier & A. Heinze (Eds.), *Proceedings of the 37th*

Conference of the International Group for the Psychology of Mathematics Education (Vol. 5, p. 106). Kiel, Germany: PME.
Lindmeier, A. (2011). *Modeling and measuring knowledge and competencies of teachers: A threefold domain-specific structure model for mathematics*. Münster, Germany: Waxmann.
Lintorf, K., McElvany, N., Rjosk, C., Schroeder, S., Baumert, J., Schnotz, W., … Ullrich, M. (2011). Zuverlässigkeit von diagnostischen Lehrerurteilen. *Unterrichtswissenschaft, 2*, 102–120.
Magnusson, S., Krajcik, L., & Borko, H. (1999). Nature, sources and development of pedagogical content knowledge. In J. Gess-Newsome (Ed.), *Examining pedagogical content knowledge* (pp. 95–132). Dordrecht, The Netherlands: Kluwer Academic publishers.
Messick, S. (1995). Validity of psychological assessment: Validation of inferences from Persons' responses and performances as scientific inquiry into score meaning. *American Psychologist, 50*(9), 741–749.
Morris, A. K., Hiebert, J., & Spitzer, S. M. (2009). Mathematical knowledge for teaching in planning and evaluating instruction: What can preservice teachers learn? *Journal for Research in Mathematics Education, 40*(5), 491–529.
Nathan, M. J., & Koedinger, K. R. (2000). An investigation of teachers' beliefs of students' algebra development. *Cognition and Instruction, 18*(2), 209–237.
Nathan, M. J., & Petrosino, A. (2003). Expert blind spot among preservice teachers. *American Educational Research Journal, 40*(4), 905–928.
Nickerson, R. S. (1999). How we know-and sometimes misjudge-what others know: Imputing One's own knowledge to others. *Psychological Bulletin, 125*(6), 737.
Nitko, A. J. (2004). *Educational assessment of students* (4th ed.). Upper Saddle River, NJ: Pearson.
van Ophuysen, S. (2010). Professionelle pädagogisch-diagnostische Kompetenz - eine theoretische und empirische Annäherung. In N. Berkemeyer, W. Bos, H. G. Holtappeis, N. McElvany, & R. Schulz-Zander (Eds.), *Jahrbuch der Schulentwicklung. Daten, Beispiele und Perspektiven* (Vol. 16, pp. 203–234). Weinheim, Germany: Juventa.
Ostermann, A., Leuders, T., & Nückles, M. (2015). Wissen, was Schülerinnen und Schülern schwer fällt. Welche Faktoren beeinflussen die Schwierigkeitseinschätzung von Mathematikaufgaben? *Journal für Mathematik-Didaktik, 36*(1), 45–76.
Ostermann, A., Leuders, T., & Nückles, M. (2017). Improving the judgment of task difficulties: Prospective teachers' diagnostic competence in the area of functions and graphs. *Journal of Mathematics Teacher Education*. https://doi.org/10.1007/s10857-017-9369-z
Park, S., & Oliver, J. S. (2008). Revisiting the conceptualisation of pedagogical content knowledge (PCK): PCK as a conceptual tool to understand teachers as professionals. *Research in Science Education, 38*(3), 261–284.
Philipp, K., & Leuders, T. (2014). Diagnostic competences of mathematics teachers – Processes and resources. *Proceedings of the 38th Conference of the International Group for the Psychology of Mathematics Education, Vol. 4*. Vancouver.
Pietsch, M. (2010). Evaluation von Unterrichtsstandards. *Zeitschrift für Erziehungswissenschaft, 13*(1), 121–148.
Popham, W. J. (2009). Assessment literacy for teachers: Faddish or fundamental? *Theory Into Practice, 48*(1), 4–11.
Praetorius, A.-K., Karst, K., & Lipowsky, F. (2012). Diagnostische Kompetenz von Lehrkräften: Aktueller Forschungsstand, unterrichtspraktische Umsetzbarkeit und Bedeutung für den Unterricht. In A. Ittel & R. Lazarides (Eds.), *Differenzierung im mathematisch-naturwissenschaftlichen Unterricht - Implikationen für Theorie und Praxis* (pp. 115–146). Bad Heilbrunn: Klinkhardt.
Prediger, S. (2010). How to develop mathematics-for-teaching and for understanding: The case of meanings of the equal sign. *Journal of Mathematics Teacher Education, 13*(1), 73–93.
Rausch, T., Karing, C., Dörfler, T., & Artelt, C. (2015). Personality similarity between teachers and their students influences teacher judgement of student achievement. *Educational Psychology: An International Journal of Experimental Educational Psychology, 36*(5), 863–878.

Rothenbusch, S. (2017). *Elementary School Teachers' Beliefs and Judgments About Students' Giftedness*. Doctoral Thesis, University Tübingen. http://dx.doi.org/10.15496/publikation-16125

Rowland, T., & Ruthven, K. (2011). Introduction: Mathematical knowledge in teaching. In T. Rowland & K. Ruthven (Eds.), *Mathematical knowledge in teaching* (pp. 1–5). Dordrecht, The Netherlands: Springer.

Santagata, R., & Yeh, C. (2016). The role of perception, interpretation, and decision making in the development of beginning teachers' competence. *ZDM, 48*(1), 153–165.

Schoenfeld, A. H. (2010). *How we think: A theory of goal-oriented decision making and its educational applications*. New York, NY: Routledge.

Schoenfeld, A. H. (2011). Noticing matters. A lot. Now what? In M. G. Sherin, V. R. Jacobs, & R. A. Philipp (Eds.), *Mathematics teacher noticing: Seeing through teachers' eyes* (pp. 223–238). New York, NY: Routledge.

Schrader, F.-W. (1989). *Diagnostische Kompetenzen von Lehrern und ihre Bedeutung für die Gestaltung und Effektivität des Unterrichts*. Frankfurt am Main, Germany: Peter Lang.

Schrader, F.-W. (2006). Diagnostische Kompetenz von Eltern und Lehrern. In D. H. Rost (Ed.), *Handwörterbuch Pädagogische Psychologie* (pp. 95–100). Weinheim, Germany: Beltz.

Schrader, F.-W. (2011). Lehrer als Diagnostiker. In E. Terhart, H. Bennewitz, & M. Rothland (Eds.), *Handbuch der Forschung zum Lehrerberuf* (pp. 683–698). Münster, Germany: Waxmann.

Schrader, F.-W. (2013). Diagnostische Kompetenz von Lehrpersonen. *Beiträge zur Lehrerinnen- und Lehrerbildung, 31*, 154–165.

Schrader, F.-W., & Helmke, A. (1987). Diagnostische Kompetenz von Lehrern: Komponenten und Wirkungen. *Empirische Pädagogik, 1*, 27–52.

Shavelson, R. J. (1978). Teacher's estimates of student 'states of mind' and behavior. *Journal of Teacher Education, 29*(5), 37–40.

Shavelson, R. J. (2010). On the measurement of competency. *Empirical Research in Vocational Education and Training, 1*, 43–65.

Shavelson, R. J., & Stern, P. (1981). Research on teachers' pedagogical thoughts, judgments, decisions, and behavior. *Review of Educational Research, 51*(4), 455–498.

Sherin, M. G., Linsenmeier, K. A., & van Es, E. A. (2009). Selecting video clips to promote mathematics teachers' discussion of student thinking. *Journal of Teacher Education, 60*, 213–230.

Sherin, M. G., & van Es, E. A. (2009). Effects of video Club participation on Teachers' professional vision. *Journal of Teacher Education, 60*(1), 20–37.

Shulman, L. S. (1986). Those who understand: Knowledge growth in teaching. *Educational Researcher*, 4–14.

Shulman, L. S. (1987). Knowledge and teaching: Foundations of the new reform. *Harvard Educational Review, 57*, 1–21.

Sloman, S. A. (1996). The empirical case for two systems of reasoning. *Psychological Bulletin, 119*, 3–22.

Spencer, L., & Spencer, S. (1993). *Competence at work*. New York, NY: Wiley.

Spinath, B. (2005). Akkuratheit der Einschätzung von Schülermerkmalen durch Lehrer und das Konstrukt der diagnostischen Kompetenz: Accuracy of Teacher Judgments on Student Characteristics and the Construct of Diagnostic Competence. *Zeitschrift für Pädagogische Psychologie, 19*(1/2), 85–95.

Streit, C., Rüede, C. & Weber, C. (2015). Diagnostische Kompetenz - Wie sich Experten und Novizen bei der Beurteilung von Schülerdokumenten unterscheiden. In F. Caluori, H. Linneweber-Lammerskitten & C. Streit (Eds.), Beiträge zum Mathematikunterricht (S. 896–899). Münster, Germany: WTM.

Südkamp, A., Kaiser, J., & Möller, J. (2012). Accuracy of teachers' judgments of students' academic achievement: A meta-analysis. *Journal of Educational Psychology, 104*(3), 743–762.

Südkamp, A., & Möller, J. (2009). Referenzgruppeneffekte im Simulierten Klassenraum: direkte und indirekte Einschätzungen von Schülerleistungen. *Zeitschrift für Pädagogische Psychologie, 23*, 161–174.

Swets, J. A., Dawes, R. M., & Monahan, J. (2000). Psychological science can improve diagnostic decisions. *Psychological Science in the Public Interest*, *1*(1), 1–26.
Tversky, A., & Kahneman, D. (1974). Judgments under uncertainty. Heuristic and biases. *Science*, *185*, 1124–1131.
Weinert, F. E. (1998). Vermittlung von Schlüsselqualifikationen. In S. Matalik & D. Schade (Eds.), *Entwicklungen in Aus- und Weiterbildung. Anforderungen, Ziele, Konzepte* (pp. 23–43). Baden-Baden, Germany: Nomos, S.
Weinert, F. E. (2001). Concept of competence: A conceptual clarification. In D. S. Rychen & L. H. Salganik (Eds.), *Defining and selecting key competencies* (pp. 45–66). Göttingen, Germany: Hogrefe.
Westera, W. (2001). Competences in education: A confusion of tongues. *Journal of Curriculum Studies*, *33*(1), 75–88.

Diagnostic Skills of Mathematics Teachers in the COACTIV Study

Karin Binder, Stefan Krauss, Sven Hilbert, Martin Brunner, Yvonne Anders, and Mareike Kunter

In the present chapter we introduce theoretical and empirical approaches to the construct of diagnostic competence within the COACTIV research program. We report eight conceptualizations and operationalizations of diagnostics skills and add in addition three constructs that seem to be close to diagnostic skills. Correlational analyses reveal only moderate and unsystematic relationships. However, our analyses showed the expected differences between school types. The chapter concludes with structural equation models in which the predictive validity of diagnostic skills for mathematical achievement of students is analyzed (both with black box and with mediation models).

1 Introduction

Diagnostic skills of teachers are – among other competence aspects such as professional knowledge, certain beliefs, motivational orientation or self-regulation – considered to be relevant both for planning lessons and for teaching. In the COACTIV study (Cognitive Activation in the Mathematics Classroom and Professional Competence of Teachers; Kunter, Baumert et al., 2013) various facets of these competence aspects of mathematics teachers were assessed including several facets of

K. Binder (✉) • S. Krauss • S. Hilbert
University of Regensburg, Regensburg, Germany
e-mail: Karin.Binder@mathematik.uni-regensburg.de

M. Brunner
University of Potsdam, Potsdam, Germany

Y. Anders
Free University of Berlin, Berlin, Germany

M. Kunter
Goethe University Frankfurt, Frankfurt, Germany

T. Leuders et al. (eds.), *Diagnostic Competence of Mathematics Teachers*, Mathematics Teacher Education 11, DOI 10.1007/978-3-319-66327-2_2

diagnostic skills. In this chapter, we review the COACTIV results with respect to diagnostic competence published so far (e.g., Anders, Kunter, Brunner, Krauss, & Baumert, 2010; Brunner, Anders, Hachfeld, & Krauss, 2013; Krauss & Brunner, 2011) and add new analyses on: (1) relationships between the different aspects of diagnostic skills, (2) respective school type differences and (3) the predictive validity for teachers' lesson quality and student learning.

As the theoretical framework on diagnostic competence is introduced in several chapters of the present publication (e.g., Leuders, T., Dörfler, Leuders, J., & Philipp, 2018; see also Südkamp & Praetorius, 2017), we will – after a short introduction into the COACTIV research program as a whole – concentrate on conceptualizations and operationalization of the related constructs in the COACTIV study.

2 The COACTIV Framework

The German COACTIV 2003/2004 research program (Cognitive Activation in the Mathematics Classroom and Professional Competence of Teachers) empirically examined a large representative sample of secondary mathematics teachers whose classes participated in the German PISA study and its longitudinal extension during 2003/04 (for an overview, see the COACTIV compendium by Kunter, Baumert et al., 2013; for PISA 2003/2004, see Prenzel et al., 2004). The structural combination of the two large scale studies PISA and COACTIV offered a unique opportunity to collect a broad range of data about students and their teachers, and to address the connection of teacher characteristics with their lesson quality and with their students' achievement (e.g., Kunter, Klusmann et al., 2013, also see Fig. 6).

In the teacher competence model of COACTIV (Fig. 1) the overarching competence aspects are: *professional knowledge, professional beliefs, motivational orientations* and *professional self-regulation skills* (the model is explicated in detail in Baumert & Kunter, 2013). In order to empirically examine research questions regarding these competence aspects (e.g., concerning their structure, school-type differences, or their impact on student learning), one needs valid and reliable measurement instruments. In COACTIV, a variety of such instruments were developed and implemented with the sample of the "COACTIV-teachers", who taught the grade 9/10 students assessed by the PISA study in 2003/2004.

In Fig. 1 (taken from Brunner et al., 2013), diagnostic skills do not appear as an autonomous competence or knowledge domain. Instead they were allocated at the intersection of pedagogical content knowledge (PCK) and pedagogical knowledge (PK). This allocation, however, remained theoretical in nature because to date no correlations of specific diagnostic skills with teacher's PCK or aspects of PK within the COACTIV data have been reported.

When analyzing such relationships in the following, we will, in addition to diagnostic skills concerning *cognitive* dimensions (such as judging student abilities or task difficulties), also include a teacher scale of diagnostic skills with respect to *social* issues ("DSS") in the following. Furthermore, in the present chapter we will

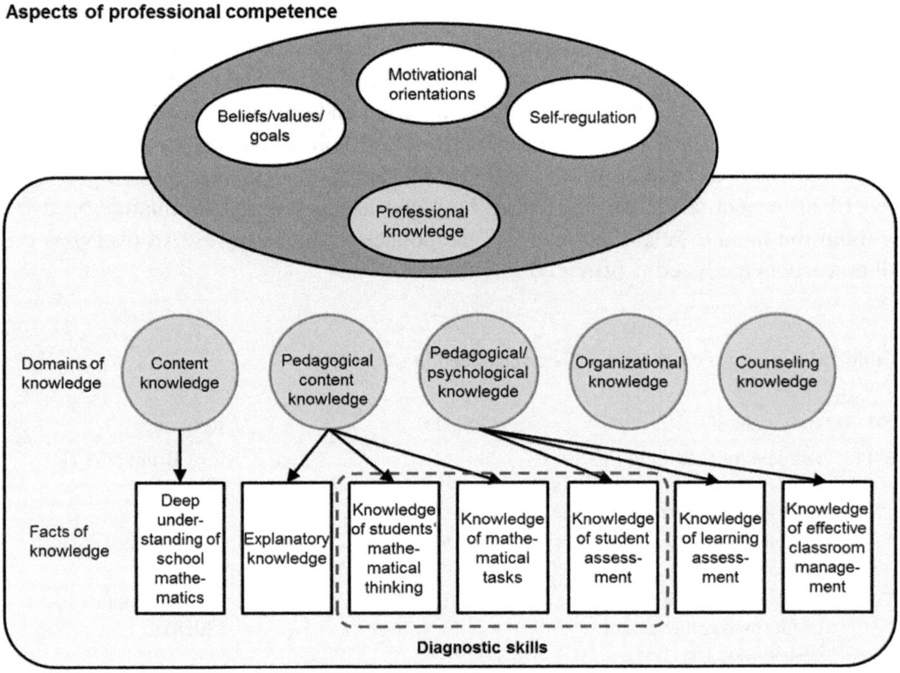

Fig. 1 The COACTIV teacher competence model (Brunner et al., 2013) and the theoretical allocation of diagnostic skills (*dashed line*)

introduce another competence aspect of mathematics teachers, which is close to diagnostic skills, namely the ability to quickly judge student answers as mathematically correct or false – which we call *quick judgment skills* (QJS)[1]. It should be noted that because previous analyses based on the COACTIV data yielded only moderate correlations between different aspects of diagnostic competence (Anders et al., 2010; Brunner et al. 2013), the COACTIV research group decided to preferably use the term *diagnostic skills*.

2.1 Diagnostic Skills Assessed in COACTIV

In contrast to the domains of pedagogical content knowledge (PCK) and content knowledge (CK), both of which in COACTIV were measured by means of open test items (Krauss et al., 2013), the eight diagnostic skills, which we introduce in Sect. 2.1.1, were assessed by relating a teacher judgment on student achievement (or motivation) to the actual PISA data of his/her students. In the model of Leuders

[1] To date, this competence was described only in a German publication and not in the framework of diagnostic skills (Krauss & Brunner, 2011).

et al. (2018), all these skills (except PCK and CK) pertain to "diagnostic thinking" (Fig. 5), because they all involve perception and interpretation of the mathematical competence of both the own class as a whole and of individual students.

In Sect. 2.1.2 we describe the conceptualization and operationalization of teacher characteristics that are related to diagnostic skills such as the skill to quickly judge student answers as correct or false (QJS), PCK and DSS. The teacher competences described in Sect. 2.1.2 were assessed by paper and pencil questionnaires or tests, without the need to relate the teachers' responses to their classes. An overview on all constructs analyzed is provided in Table 1a.

Table 1a Aspects of diagnostic skills investigated in COACTIV

Construct D1–D8 judgment of …		Scale	Reference
D1	Achievement level in PISA test (class average) compared to school type specific German average	1–5[a]	e.g., Brunner et al. (2013)
D2	Distribution of achievement (in class)	1–5[a]	e.g., Brunner et al. (2013)
D3	% of own students in bottom third of German achievement distribution (in class)	0–100%	e.g., Brunner et al. (2013)
D4	% of own students in top third of German achievement distribution (in class)	0–100%	e.g., Brunner et al. (2013)
D5	Motivational level (class average) compared to school type specific average	1–5[a]	e.g., Brunner et al. (2013)
D6	% correct solutions with respect to four specific PISA tasks (in class)	0–100% (for each task)	e.g., Anders et al. (2010), Brunner et al. (2013) (see Fig. 2)
D7	Solutions of two specific tasks (Kite and Mrs. May) with respect to seven specific students	2 × 7: Yes/no	e.g., Brunner et al. (2013) (see Figs. 2 and 3 left)
D8	Rank order of achievement of seven specific students in PISA (diagnostic sensitivity)	Distribution of position numbers 1–7	e.g., Anders et al. (2010), Brunner et al. (2013) (see Fig. 3 right)
QJS	Quickly classifying student answers as correct or false (12 tasks provided with respective student responses)	# correct judgments divided by the mean of 12 reaction times	e.g., Krauss and Brunner (2011) (see Fig. 4)
PCK	Pedagogical content knowledge	22 open test items	e.g., Krauss et al. (2008, 2013)
DSS	Diagnostic skills concerning social issues	4 items, each 1–4[b]	Not yet published

[a]1 = "considerably below average", 2 = "somewhat below average", 3 = "average", 4 = "somewhat above average", 5 = "considerably above average"
[b]1 = "strongly disagree", 2 = "disagree", 3 = "agree", 4 = "strongly agree"

2.1.1 Diagnostic Skills (D1–D8)

In COACTIV several established instruments (Hoge & Coladarci, 1989; McElvany et al., 2009; Schrader, 1989) were implemented, targeting different objects of judgment (student achievement vs. motivation; performance on specific tasks vs. the full PISA test) and different levels of judgment (individual students vs. whole class). In the following we will describe the operationalization of eight measures of diagnostic skills[2] ("D1–D8", see Table 1a).

At the class level, teachers were asked to provide the following ratings: "Please rate the *achievement level* of your PISA class in mathematics relative to an average class of the same school type" **(D1)**; "Please rate the *distribution of achievement* in mathematics in your PISA class relative to an average class of the same school type" **(D2)**; and "Please rate the *motivation* of your PISA class in mathematics relative to an average class of the same school type" **(D5)**. All responses were given on a five-point rating scale with the options "considerably below average" (coded 1), "somewhat below average" (coded 2), "average" (coded 3), "somewhat above average" (coded 4) and "considerably above average" (coded 5). To determine the accuracy of the teachers' judgments, teacher responses then were compared with the actual outcomes of their PISA classes. As it is common in the literature on diagnostic competence, small judgment errors (i.e., absolute values of differences) were considered indicators of high diagnostic skills. To this end, we first calculated quantiles for achievement level, distribution of achievement, and motivation separately for each school type. Each PISA class was then assigned to one of these quintiles (see Spinath, 2005, for an analogous procedure). The first quintile was coded 1, the second quintile was coded 2, etc. In a second step, we computed the difference between the teachers' ratings and these objective quintiles, terming the absolute value of this difference the *judgment error* (see Table 1b). Thus, the maximal error was four and a judgment error of zero indicated that the teacher rating was congruent with the objective outcome (the detailed statistical procedure is explicated in Brunner et al., 2013).

To provide further indicators of diagnostic skills at the class level, teachers were asked to estimate the percentages of high- and low-achieving students in their PISA class by answering the following questions: "Relative to other classes of the same grade and school type, please estimate the percentage of students in your PISA class performing at a *low-achievement level* (in the bottom third)" **(D3)** and "Relative to other classes of the same grade and school type, please estimate the percentage of students in your PISA class performing at a *high-achievement level* (in the top third)" **(D4)**. To gauge the accuracy of these judgments, we then computed the judgment error in terms of the absolute difference between the teachers' judgments and the actual percentage of high- versus low-achieving students in the class (see Table 1b).

To evaluate the accuracy of teachers' assessment of task demands, we asked them to estimate how many students in their class would be able to solve each of four specific PISA tasks correctly (A, B, Ca and Cb, see Fig. 2) that addressed important domains of mathematical content typically covered at secondary level

[2] Parts of Sect. 2.1.1 are taken from Brunner et al. (2013).

Table 1b Descriptive results on the aspects of diagnostic skills. For D1–D6: Error = 0 signifies "maximal" diagnostic skill. For D7, D8, QJS, PCK and DSS positive values indicate high performance

Construct E-D1 – E-D6 judgment error of …		Scale	M (SD)
E-D1	Achievement level in PISA test (class average) compared to school type specific German average	0–4[a]	1.20 (0.86)
E-D2	Distribution of achievement (in class)	0–4[a]	1.14 (0.91)
E-D3	% of own students in bottom third of German achievement distribution (in class)	0–100%	0.14 (0.12)
E-D4	% of own students in top third of German achievement distribution (in class)	0–100%	0.22 (0.15)
E-D5	Motivational level (class average) compared to school type specific German average	0–4[a]	1.28 (0.95)
E-D6	% correct solutions with respect to four specific PISA tasks (in class) (task related jugement error)	0–100% (mean error across four tasks)	0.28 (0.11)
D7	Solutions of two specific tasks (kite and Mrs. may) with respect to seven specific students	0–100% (proportion of correct predictions of $2 \times 7 = 14$ predictions)	0.50 (0.15)
D8	Rank order of achievement of seven specific students in PISA (diagnostic sensitivity)	−1 to 1[b]	0.38 (0.35)
QJS	Quickly classifying student answers as correct or false (12 tasks provided with respective student responses)	0.20–2.38	1.02 (0.36) 12 items $\alpha = 0.71$
PCK	Pedagogical content knowledge	0–35	20.38 (5.71) 22 items $\alpha = 0.78$
DSS	Diagnostic skills concerning social issues	0–4	2.95 (0.47) 4 items $\alpha = 0.88$

Note: M mean, *SD* standard deviation
[a]Estimation within the correct quintile: judgment error of zero. Estimating, for example, the highest quintile, although the lowest quintile would be correct (or vice versa): error of four
[b]Correlation coefficient for ranking (Spearman) between estimated rank order and actual rank order of the seven students

(D6). For each task, we computed the (absolute value of) the difference between the teachers' estimates and the actual proportion of correct answers in the class as a measure of judgment error. The mean judgment error across the four tasks – the *task-related judgment error* – was then calculated (Table 1b). A task-related judgment error of zero again indicates that a teacher correctly estimated the number of correct solutions in his/her PISA class on all four tasks.

All of the above indicators relate to the class as a whole. To examine the teachers' ability to predict the performance of individual students, we additionally asked

A. "Kite"

Some students want to make kites. Peter and Rosie prepare frames out of light wooden sticks.

Then they want to stick a thin sheet of plastic film onto this frame. It has to be a single piece of film.

What is the surface area of the plastic film to be stuck on the kite?

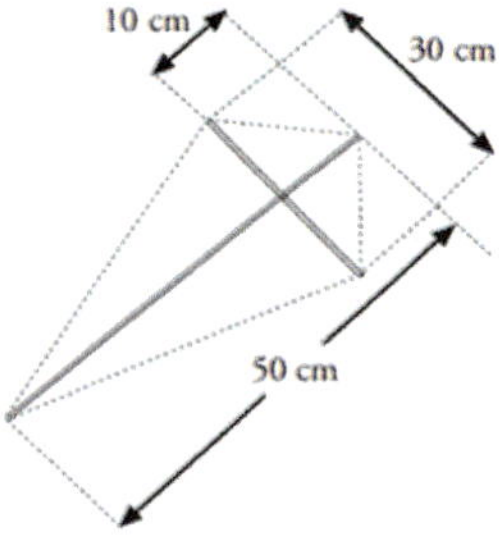

(Drawing not to scale)

B. "Mrs. May"

Mrs. May runs a clothes shop. She pays a wholesale price of €150 for a dress from a supplier.

She calculates the retail price to be written on the price tag as follows: First she increases the wholesale price by 100 %. Then she adds 16 % tax to this new price.

What price does Mrs. May write on the price tag?

C. "Sausage Stand a and b"

A class is running a sausage stand at a school fete. One student prepares a price table for bigger orders. But he makes a mistake in his calculations.

a) *Put a cross in the column containing the mistake.*

Number of sausages	3	4	6	8	10
Price	€ 3.60	€ 4.80	€ 7.20	€ 8.60	€ 12.00
	☐	☐	☐	☐	☐

b) *Give reasons for your decision and correct the mistake.*

Fig. 2 Specific PISA tasks used in COACTIV to assess teachers' diagnostic skills (i.e., D6 and D7). All four tasks were provided to the teachers

the teachers to consider seven specific students, who were drawn at random from their class. First, they rated whether or not these students would be able to solve the tasks "Kite" and "Mrs. May" correctly (see Fig. 3). The *accuracy of these individual teacher judgments* **(D7)** was determined by calculating the proportion of the 14 predictions (m = 2 tasks and n = 7 students) that were correct. The theoretically possible range was thus from 0 to 1, with a score of 1 indicating that all 14 of a teacher's predictions were correct.

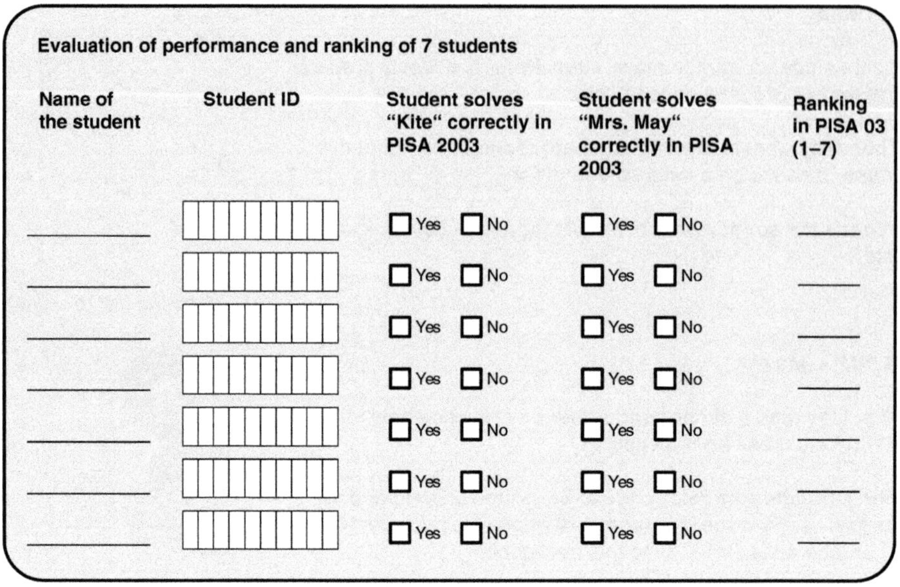

Fig. 3 Estimating the performance of seven students in two specific PISA tasks (D7) and estimated ranking of performance of these seven students in the whole PISA test (D8)

Finally, we asked the teachers to judge how well the same seven students probably performed on the whole PISA 2003 mathematics assessment by putting them in rank order of achievement (**D8**, see also Fig. 3). This estimated rank order again was compared with the students' actual PISA rank order. To provide a measure of *diagnostic sensitivity*, we then computed the rank correlation (Spearman's Rho) of the two rank orders. The higher the diagnostic sensitivity score, the better able a teacher was to predict the rank order of achievement; a score of 1 indicates a perfect prediction.

Thus, the scales of D1–D6 (Table 1a) first had to be transformed into errors E-D1 to E-D6 (Table 1b) and therefore here zero denotes maximal performance, whereas D7 and D8 refer to accuracy or sensitivity itself and therefore here positive values indicate better skills.

2.1.2 Competence Aspects Related to Diagnostic Skills

The following further constructs assessed in COACTIV (an overview on all constructs is provided in the COACTIV-scale documentation, Baumert et al., 2009) are theoretically close to diagnostic skills.

Quick Judgment Skill (QJS)

A mathematics teacher should be able to establish the truth or falsehood of students' statements (or responses to tasks) in mathematics lessons within a reasonable time. Such judgments challenge teachers' content-specific expertise, because they happen in the publicity of the classroom with all its spontaneity. As mathematics experts teachers should notice failures and they should not take too much time for their identification.

To model this time pressure, in a computer-based instrument 12 easy mathematical tasks were implemented, each with a (hypothetical) corresponding student answer (for a screenshot see Fig. 4). The instruction for the COACTIV teachers was for each task to judge as quickly as possible whether the provided student response was correct or false (all tasks and respective student answers should – without time pressure – constitute no problems for mathematics teachers; all tasks are listed in Krauss & Brunner, 2011). When task and respective student answer were presented simultaneously, the time began to run.

For each teacher and each task the correctness and the time needed for the judgment were recorded. The score for the QJS of a teacher then was calculated by dividing the number of correct judgments (out of 12) by the mean reaction time.

Pedagogical Content Knowledge (PCK)

In COACTIV, two paper and pencil tests on the pedagogical content knowledge (PCK) and the content knowledge (CK) of mathematics teachers were administered (Krauss et al., 2008). Especially the PCK test is of relevance with respect to the present chapter, since two out of three knowledge facets addressed in this test relate

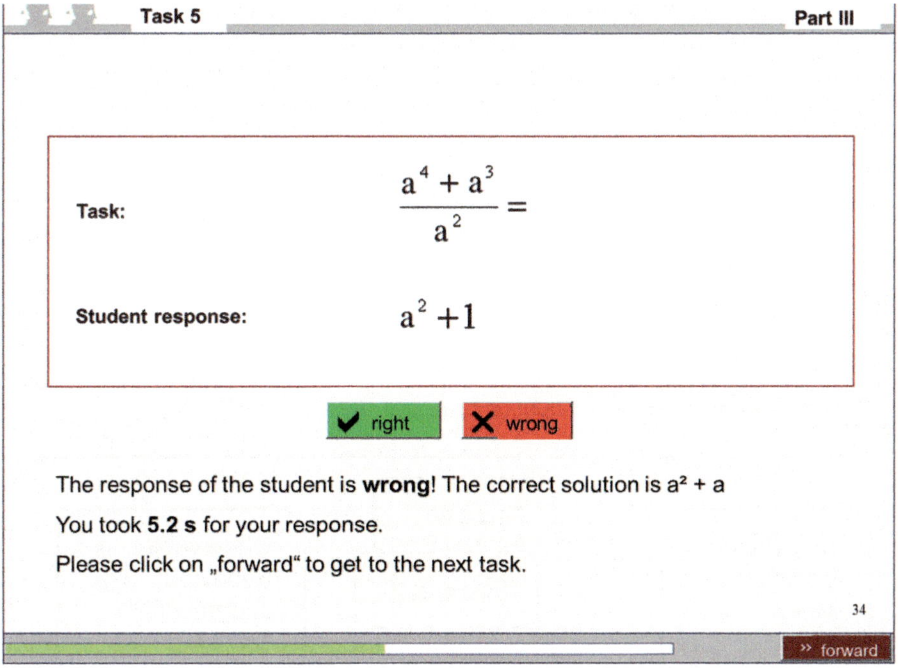

Fig. 4 Sample item of the reaction time test used in COACTIV to assess teachers' QJS (the three lines below appeared after the teacher made his/her decision)

to diagnostic skills. Altogether, PCK was operationalized by 22 items on (for details on the tests and respective results see Krauss et al., 2013):

- Explaining and representing mathematical contents (11 items)
- Mathematics-related student cognitions (typical error and difficulties, 7 items)
- The potential of mathematical tasks (for multiple solution paths, 4 items)

The latter two aspects are closely related to diagnostic skills of mathematics teachers, since they refer to students' thinking and to task properties. Yet – in contrast to D1–D8 – in the PCK test teachers had to answer general questions on typical students' mathematical difficulties and on task properties (such as: "Which problems students *typically* face when …"), i.e., there was no need to relate the answers to actual PISA data.

Diagnostic skills Concerning Social Issues
Diagnostic skills with respect to social issues are another interesting competence facet that might be related to the content-specific skills described so far. In COACTIV, the teacher-scale "diagnostic skills concerning social issues" consists of four items. One sample item was "I notice very quickly, if someone is really sad".

2.1.3 General Model of Diagnostic Competence

Looking at diagnostic skills through various different glasses as in COACTIV is in line with the call of Südkamp and Praetorius (2017), to assess diagnostic competence not in a narrow and constrained sense, but with multiple measures. According to the model of Leuders et al. (2018), which is close to the model of Südkamp and Praetorius (2017), PCK, QJS and DSS are considered diagnostic dispositions, because they were assessed by paper and pencil questionnaires or tests in a laboratory setting outside of the classroom. In contrast, D1–D8 clearly require teachers' perception and interpretation of aspects of their real classes and then to come up with a decision. Yet, because actual decisions were not observed in the real classroom context (but again in the laboratory setting), we theoretically subsume D1–D8 under diagnostic thinking (middle column of Fig. 5).

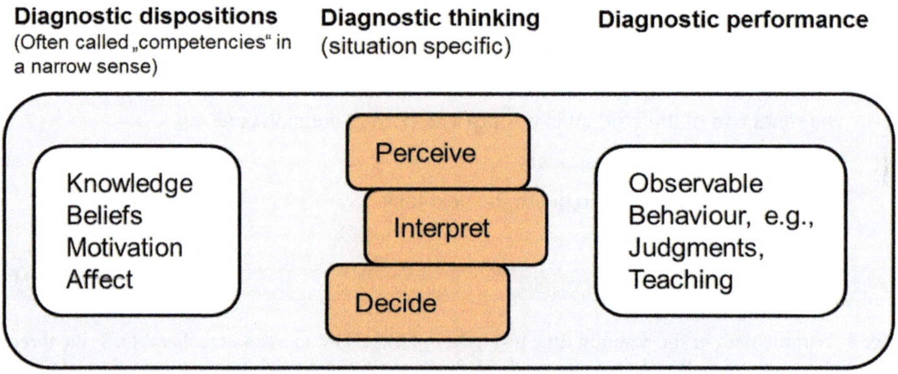

Fig. 5 Diagnostic competence (in a wider sense) as a continuum (Leuders et al., 2018)

2.2 *Instructional Quality and Student Characteristics*

In order to assess teachers' instructional quality in COACTIV, a parsimonious model with three latent dimensions, which are each represented by multiple indicators, was developed (Fig. 6, for details see Kunter and Voss, 2013). Very briefly, the potential for *cognitive activation* was assessed in terms of the cognitive quality of the mathematical tasks implemented by the teachers in class tests (e.g., the need for mathematical argumentation; cf. Kunter et al., 2013). Class tests were chosen because they allow valid conclusions to be drawn about the intended purposes of instruction. The dimension of *classroom management* was assessed using scales from both the student (PISA) and the teacher (COACTIV) questionnaires asking, for instance, for disruption levels or time wasted. Indicators of *individual learning support* were formed by scales from the student questionnaire, assessing various aspects of the interaction between students and teachers (see Kunter & Voss, 2013). The students' learning gain was estimated by the mathematical achievement in PISA 2004 (while controlling for the achievement in the preceding year, i.e. PISA 2003). The full mediation model is explicated in Figs. 6 and 8.

In Sect. 3.4 structural equation modeling will be conducted in order to estimate the predictive validity of aspects of diagnostic skills for lesson quality and students' learning gains (structural equation models with respect to various other teacher competence aspects as predictors are summarized in Kunter, Baumert et al., 2013, Kunter, Klusmann et al., 2013, or in Krauss et al., 2017).

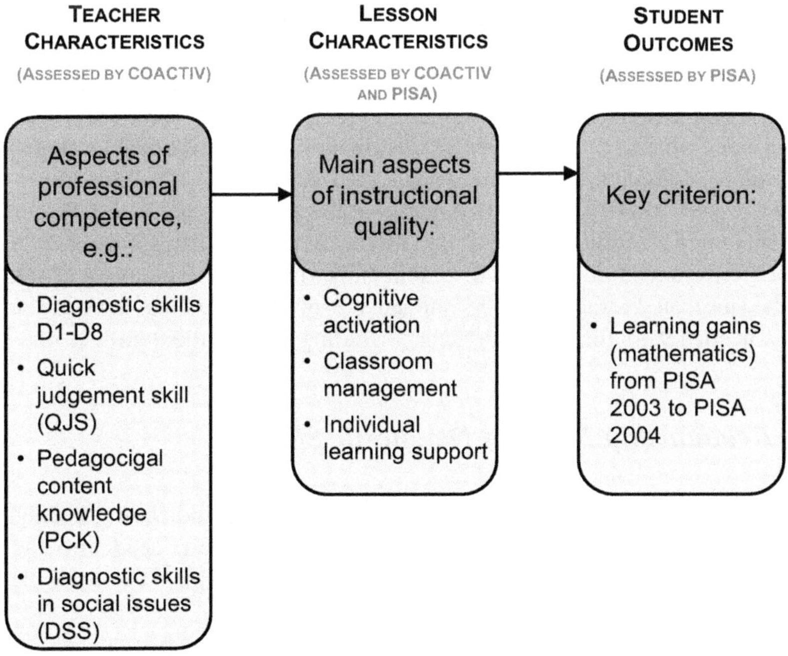

Fig. 6 Causal model in COACTIV

So far, empirical evidence on the predictive validity of aspects of diagnostic competence for student learning is mixed (see e.g., Gabriele, Joram, & Park, 2016). Several studies document predictive validity, some studies show moderation by instructional variables and other studies find no predictive evidence. It can be assumed that the precise operationalization and measurement of diagnostic competence are essential and that the respective variation might explain the differing empirical results at least partially (e.g., Artelt & Rausch, 2014).

3 Results

In Sect. 3.1, we report descriptive results with respect to the constructs assessed and in Sect. 3.2 we analyze the relationship between the different diagnostic skills. As previous research has shown major differences between teachers' competences with respect to different German secondary school types (Kunter, Baumert et al., 2013), we also report respective differences in diagnostic skills in Sect. 3.3. Finally, in Sect. 3.4, we conduct structural equation modeling in order to analyze the effects of diagnostic skills on instructional quality and on learning gains of students.

3.1 Descriptive Results

To estimate teachers' diagnostic skills with respect to the constructs displayed in Table 1a, judgment errors regarding the skills D1–D6 were calculated (Table 1b). While D1–D5 and D8 depend on single item measures, QJS, PCK and DSS consist of multi-item-scales (therefore for the latter three constructs, Cronbach's alpha is provided in Table 1b). Since the scales of E-D1, E-D2 and E-D5 share the same range, judging achievement level, achievement distribution and motivational level obviously was similarly difficult (yielding means of 1.18, 1.22 and 1.30, see Table 1b).

With respect to D7 it should be noted that the empirical mean almost perfectly represents the probability of guessing. Obviously it is difficult for teachers to predict the performance of individual students in certain tasks. The mean of QJS was 1.02, because on average teachers judged 8.8 of the 12 items correctly and on average needed 9.7 s for each judgment (including reading the item).

3.2 Relationship Between Diagnostic Skills

Table 2 shows the intercorrelations between the various indicators of diagnostic skills for the whole sample (in each cell above) and for the academic and non-academic track separately (the two values in each cell below)[3]. In Germany there are

[3] The intercorrelations with QJS, PCK and DSS as well as the school type related correlations were not reported in Brunner et al. (2013).

basically three secondary school types, namely *Gymnasium* (academic track, prerequisite for the admission to university), *Realschule* (intermediate track) and *Hauptschule* (vocational track). The respective teacher education differs between Gymnasium (higher proportion of content courses) and the latter two school types (higher proportion of educational courses). Therefore in COACTIV analyses usually the performance of the academic track teachers (GY) is compared with teachers of the other tracks, which are called Non-Gymnasium (NGY). For details concerning the German school system see, for example, Cortina and Thames (2013).

In Table 2, E-D1 to E-D6 denote errors and D7 to DSS denote the competences themselves (displayed within the dottet rectangle in Table 2). Thus, theoretically the correlations of E-D1 to E-D6 and of D7 to DSS should be positive, while each of the errors should correlate negatively with each of the competences.

As it becomes clear from Table 2, there seems to be no systematic pattern within the bivariate correlations. Because there is no appearance of a dominant dimension of diagnostic competence, the constructs D1 to D8 were named “diagnostic skills” by Brunner et al. (2013). This pattern of results – only weak or no correlations between different indicators of diagnostic skills – was also reported by both Schrader (1989) and Spinath (2005).

However, judging the achievement level (D1) and judging the bottom third (D3) and the top third of achievement distribution (D4) seem to be correlated as well as the skills D6, D7 and D8. Interestingly, DSS is even associated negatively with some other competence aspects. It seems to be that teachers, who concentrate on social aspects of students, are less competent with respect to, for example, D3, D8 and PCK. Furthermore, there are highly differential correlations of D6 and QJS: While in the group of NGY-teachers the correlations are significantly negative, in the GY-group the opposite is true.

However, the results in Table 2 should be interpreted with caution, because only about half of the correlations point in the expected direction. Of course the different measurements of diagnostic skills can be criticized. Perhaps constructing quintiles, for instance, may not be the best procedure to judge D1, D2 and D5. Furthermore, estimating students’ performance in the whole PISA test might be difficult, because teachers did not know all items of this test. However, Table 2 corroborates the assumption that measuring judgment accuracies might highly depend on the exact operationalization (Gabriele et al., 2016).

3.3 School-Type Differences with Regard to the Mean Levels

In Table 3, mean levels of the teachers of the academic track (GY) are compared with the non-academic track teachers (NGY). There were differences with respect to school type in favor of GY-teachers, especially regarding D4, D6, QJS and PCK. Interestingly, the diagnostic skills with respect to social issues are descriptively more pronounced in NGY-teachers.

Table 2 Intercorrelations of the indicators of diagnostic skills (Pearson product–moment correlation) for all teachers (upper half in each cell) and for teachers of academic (GY) and non-academic track (NGY) separately (lower half)

All (GY / NYG)	E-D1		E-D2		E-D3		E-D4		E-D5		E-D6		D7		D8		QJS		PCK		DSS	
E-D1	--																					
E-D2	0.12		--						all teachers													
	0.23	0.01							GY	NGY												
E-D3	**0.18**		-0.10		--																	
	0.06	**0.26**	-0.06	-0.12																		
E-D4	***0.29****		0.11		-0.09		--															
	0.21	***0.34***	0.18	0.08	***-0.30***	0.00																
E-D5	-0.05		-0.02		-0.08		-0.09		--													
	-0.09	-0.02	-0.12	0.06	0.03	-0.15	-0.22	-0.04														
E-D6	-0.01		0.03		0.00		0.08		0.07		--											
	0.01	-0.03	-0.01	0.06	0.01	-0.05	-0.13	0.10	0.09	0.05												
D-7	-0.04		-0.05		0.14		-0.09		-0.11		***-0.33***		--									
	-0.05	0.01	0.13	**-0.23**	0.19	0.11	0.03	-0.12	-0.16	-0.07	-0.19	***-0.40***										
D-8	0.00		0.03		-0.07		0.03		0.07		**-0.16**		**0.17**		--							
	0.01	-0.02	0.15	-0.09	0.01	-0.13	0.04	0.02	0.13	0.01	-0.23	-0.14	0.18	0.17								
QJS	0.03		0.06		-0.03		-0.01		-0.09		-0.14		0.09		0.00		--					
	0.12	-0.05	0.08	0.04	0.06	-0.07	0.02	0.05	-0.03	-0.12	0.20	**-0.27**	-0.05	0.19	-0.07	0.09						
PCK	0.06		0.10		-0.14		-0.08		-0.14		-0.08		*-0.07*		0.13		0.13		--			
	0.09	0.05	-0.01	0.19	-0.13	-0.11	0.08	-0.10	-0.20	-0.09	0.05	0.01	-0.00	-0.21	-0.04	**0.27**	-0.03	0.11				
DSS	-0.01		-0.06		**0.17**		-0.01		0.03		0.07		-0.09		**-0.17**		0.02		**-0.20**		--	
	0.13	-0.14	0.00	-0.11	**0.28**	***0.09***	-0.13	0.05	-0.08	0.13	0.14	-0.02	-0.10	-0.06	***-0.30***	-0.05	-0.02	0.09	-0.14	**-0.22**		

$N_{All, min}=136$
$N_{All, max}=180$

$N_{GY, min}=62$
$N_{GY, max}=77$

$N_{GY, min}=74$
$N_{GY, max}=103$

*__Bold__: p≤0.05, ***Bold and italics***: p≤0.01

Table 3 Differences in diagnostic skills by school type

Dimension of (sub-)skills	Academic track GY N M (SD)	Non-academic track NGY N M (SD)	Group differences	
			d	*p*-value
E-D1 achievement level in PISA-test (class average) compared to school type specific German average	*N* = 103 1.12 (0.94)	*N* = 78 1.20 (0.86)	0.09	0.51
E-D2 distribution of achievement (in class)	*N* = 103 1.13 (1.03)	*N* = 77 1.14 (0.91)	0.01	0.97
E-D3% of own students in bottom third of German achievement distribution (in class)	*N* = 102 0.13 (0.11)	*N* = 75 0.14 (0.12)	0.09	0.49
E-D4% of own students in top third of German achievement distribution (in class)	*N* = 101 0.17 (0.11)	*N* = 75 0.22 (0.15)	0.38	0.02
E-D5 motivational level (class average) compared to school type specific German average	*N* = 102 1.23 (0.87)	*N* = 77 1.28 (0.95)	0.05	0.72
E-D6% correct solutions with respect to four specific PISA tasks in class (task related judgment error)	*N* = 88 0.22 (0.08)	*N* = 66 0.28 (0.11)	0.63	<0.01
D7 solutions of two specific tasks (Kite and Mrs. May) with respect to seven specific students	*N* = 91 0.54 (0.15)	*N* = 74 0.50 (0.15)	0.27	0.10
D8 rank order of achievement of seven specific students in PISA (diagnostic sensitivity)	*N* = 91 0.37 (0.37)	*N* = 74 0.38 (0.35)	−0.03	0.93
QJS quickly classifying student answers as correct or false (12 tasks provided with respective student responses)	*N* = 87 1.13 (0.38)	*N* = 71 0.94 (0.33)	0.54	<0.01
PCK pedagogical content knowledge	*N* = 94 22.50 (5.43)	*N* = 73 18.46 (5.68)	0.72	<0.01
DSS diagnostic skills concerning social issues	*N* = 99 2.88 (0.53)	*N* = 77 2.95 (0.47)	−0.14	0.36

Note*:* *M* mean, *SD* standard deviation, *d* effect size according to Cohen (1992). The effect sizes in Table 3 were always calculated in a way, so that *positive effect sizes mean advantage of the GY-teachers* (already acknowledging that lower errors denote higher performances)

These results mirror previous COACTIV results since we also found effects in favor of GY-teachers with respect to many other *content-related* competence aspects and effects in favor of NGY-teachers regarding some further *non-content-related* competences in COACTIV (Kunter, Baumert et al., 2013).

3.4 Predictive Validity with Respect to Teaching Quality and Student Learning

Despite all problems concerning reliability and validity of the constructs D1 to D8 reported above, we ran a series of two-level structural equation models to tentatively check the predictive validity of the constructs assessed. Since the large predictive validity of PCK was previously documented in various black box and in mediation models (Baumert & Kunter, 2013; Baumert et al., 2010; Kunter, Klusmann et al., 2013), this construct will be ignored in the following (for further information on objectivity, reliability and validity of QJS or PCK, see Krauss & Brunner, 2011 or Krauss et al., 2013, respectively).

First, we specified nine separated black box models (see Fig. 7 or Table 4) where D1–D8 and QJS should predict student achievement in PISA 2004. At the class level, we controlled for school type (GY vs. NGY) and on the individual level we controlled for prior knowledge in mathematics (PISA achievement in 2003), reading literacy, basic cognitive abilities, immigration status and socio-economic status

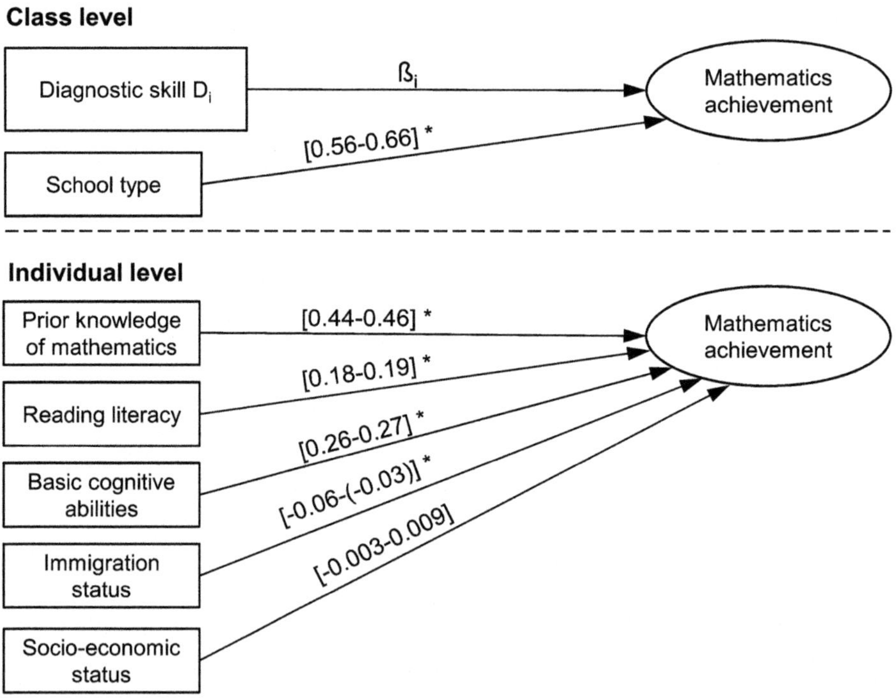

Fig. 7 Black box models with the predictors D1–D8 and QJS and the criterion students' mathematics achievement (the single results for the standardized regression coefficients β_i are depicted in Table 4); $^*p<0.05$

Table 4 Nine black box models and the respective standardized regression coefficients β for Fig. 7 (criterion: mathematics achievement)

Model	Predictor	β	p-value
Model 1 E-D1	E-D1 achievement level in PISA test (class average)	0.28*	<0.01
Model 2 E-D2	E-D2 distribution of achievement (in class)	0.03	0.81
Model 3 E-D3	E-D3% students in bottom third of achievement distribution (in class)	−0.14	0.16
Model 4 E-D4	E-D4% students in top third of achievement distribution (in class)	0.27*	<0.01
Model 5 E-D5	E-D5 motivational level (class average)	−0.03	0.77
Model 6 E-D6	E-D6% correct solutions with respect to 4 specific PISA tasks (in class)	−0.21	0.06
Model 7 D7	D7 solutions of 2 specific tasks (kite and Mrs. may) with respect to seven specific students	0.07	0.53
Model 8 D8	D8 rank order of these seven specific students in PISA	−0.02	0.83
Model 9 QJS	QJS quickly classifying student answers as correct or false (12 tasks with respective student responses)	−0.03	0.80

β Standardized regression coefficient, *$p<0.05$

(in brackets in Fig. 7 the range of the corresponding standardized regression coefficients with respect to all nine models is depicted).

Table 4 summarizes the results of all nine black box models. It should be noted that the significant positive coefficients in model 1 and model 4 denote *negative effects*, because with respect to D1 and D4 errors were modeled (for an explanatory attempt, see the respective mediation models later). The only model that is close to a *positive effect* on student achievement is model 6 (judging the proportions of correct solutions in four specific PISA tasks in the class). Anders et al. (2010) demonstrated that this effect becomes significant if only the two tasks "Sausage Stand a" and "Sausage Stand b" are analyzed (maybe because proportionality is an intensively treated topic in mathematics in Germany). We could not replicate the positive effect of D8 from Anders et al. (2010) because of a different composition of the teacher sample analyzed.

In a second series we implemented nine corresponding mediation models[4] (Fig. 8). The measurement of the instructional quality by the latent constructs cognitive activation (assessed by cognitive level of tasks), learning support and classroom management was previously described in Sect. 2.2. We found that the negative effects of D1 and D4 were mediated in both cases by a positively significant

[4] In the tradition of COACTIV we name these models "mediation models" (Baumert et al., 2010, Kunter, Klusmann et al., 2013), although the use of this term usually implies addressing, e.g., the multiplicative term of the indirect path, etc.

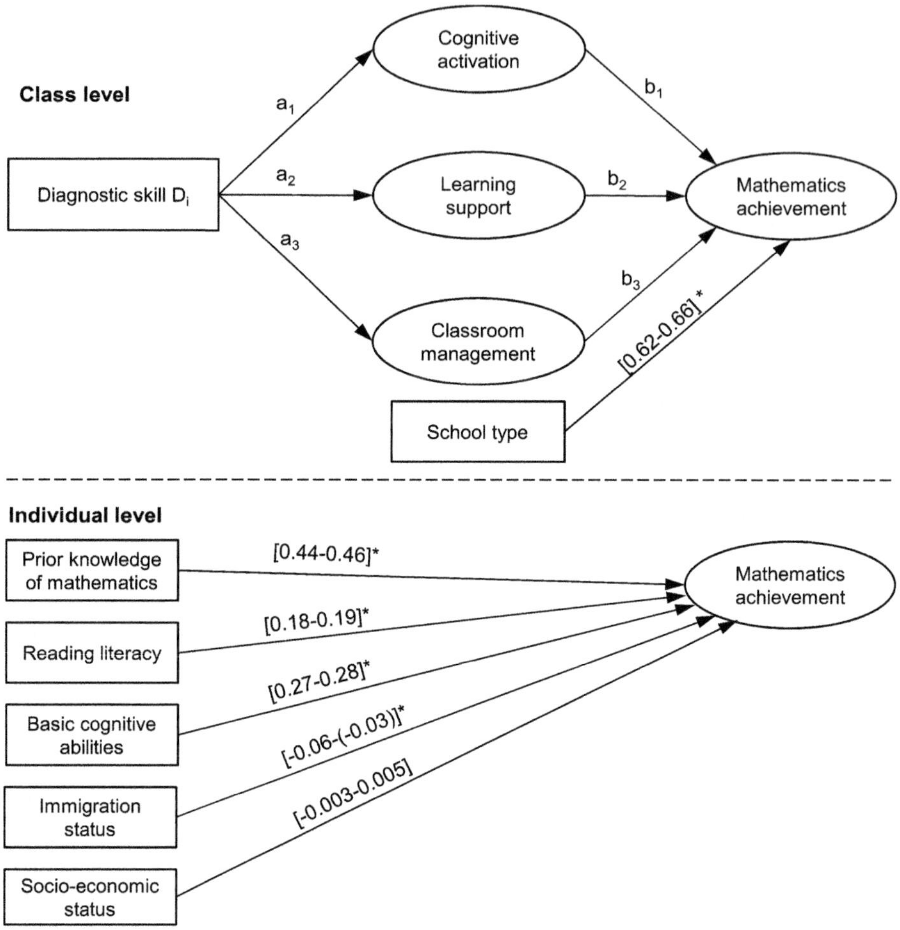

Fig. 8 Overview of the nine two-level "mediation" models (*Note*: a_i and b_i denote the respective standardized regression coefficients, D1-D8: predictors, *p<0.05)

path a_1 (see Fig. 8), which means that teachers with *less* D1 and D4 implemented *more* cognitively demanding tasks in their classes. This is an interesting finding since – *although,* or perhaps even *because* of misjudging (!) the achievement level – teachers dare to implement cognitively activating tasks, which in turn leads to higher mathematics achievement.

The only significant effect we found with respect to a_2 and a_3 was in model 8, were a_2 was positively significant and a_3 was negatively significant, indicating that D8 (estimating the rank order of seven students) works differently than the other predictors. Concerning b_1 to b_3 in almost all models b_1 and b_3, were significant while b_2 was not (which replicated the results of other structural equation models of COACTIV, see, e.g., Kunter, Klusmann et al., 2013).

4 Discussion

In the present chapter, we summarized the findings of Anders et al. (2010) and Brunner et al. (2013) and added the constructs of quickly judging student responses (QJS), pedagogical content knowledge (PCK) and diagnostic skills concerning social issues (DSS), which – at least theoretically – should be close to diagnostic skills. However, correlational analyses yielded only moderate and even partially unexpected results. In contrast, mean level differences between Gymnasium- and Non-Gymnasium-teachers are in line with previous COACTIV results and thus (cautiously) validate the constructs implemented.

Finally, we implemented structural equation models to assess the impact of diagnostic skills and QJS directly on students' mathematical achievement (black box models) and models where this assumed effect was mediated by central aspects of instructional quality. Interestingly, the precise judgment of student achievement level may even prevent teachers from implementing cognitively demanding mathematical tasks (mediation models 1 and 4). Only the correct judgment of the solution rate with respect to four specific tasks (that explicitly were shown to the teachers) seems to have a positive impact on students' achievement (black box model 6). Note that estimating the performance of their class in the whole PISA test obviously was difficult for teachers, maybe because they do not know the concrete tasks implemented in PISA.

Taken together, our analyses confirm the use of the term "diagnostic skills" (instead of diagnostic competence) by Brunner et al. (2013), because we found only unsystematic and moderate relationships between the constructs analyzed. Our results are in line with Spinath (2005), who also found only weak or no correlations between different indicators of diagnostic skills. The present chapter, however, is far from stating final conclusions, but aims to introduce and describe various ways to theoretically and empirically examine diagnostic skills of teachers in the subject of mathematics.

References

Anders, Y., Kunter, M., Brunner, M., Krauss, S., & Baumert, J. (2010). Diagnostische Fähigkeiten von Mathematiklehrkräften und ihre Auswirkungen auf die Leistungen ihrer Schülerinnen und Schüler. *Psychologie in Erziehung und Unterricht, 57*, 175–193.

Artelt, C., & Rausch, T. (2014). Accuracy of teacher judgments: When and for what reasons? In S. Krolak-Schwerdt, S. Glock, & M. Böhmer (Eds.), *Teachers' professional development: Assessment, training, and learning* (pp. 229–248). Rotterdam, The Netherlands: Sense Publishers.

Baumert, J., Blum, W., Brunner, M., Dubberke, T., Jordan, A., Klusmann, U., Krauss, S., Kunter, M., Löwen, K., Neubrand, M., & Tsai, Y.-M. (2009). *Professionswissen von Lehrkräften, kognitiv aktivierender Mathematikunterricht und die Entwicklung von mathematischer Kompetenz (COACTIV): Dokumentation der Erhebungsinstrumente* (Materialien aus der Bildungsforschung, 83). Berlin, Germany: Max-Planck-Institut für Bildungsforschung.

Baumert, J., Kunter, M., Blum, W., Brunner, M., Voss, T., Jordan, A., Klusmann, U., Krauss, S., Neubrand, M., & Tsai, Y.-M. (2010). Teachers' mathematical knowledge, cognitive activation in the classroom, and student progress. *American Educational Research Journal, 47*(1), 133–180.

Baumert, J., & Kunter, M. (2013). The COACTIV model of teachers' professional competence. In M. Kunter, J. Baumert, W. Blum, U. Klusmann, S. Krauss, & M. Neubrand (Eds.), *Cognitive activation in the mathematics classroom and professional competence of teachers, mathematics teacher education* (Vol. 8, pp. 25–48). New York, NY: Springer.

Brunner, M., Anders, Y., Hachfeld, A., & Krauss, S. (2013). The diagnostic skills of mathematics teachers. In M. Kunter, J. Baumert, W. Blum, U. Klusmann, S. Krauss, & M. Neubrand (Eds.), *Cognitive activation in the mathematics classroom and professional competence of teachers mathematics teacher education* (Vol. 8, pp. 229–248). New York, NY: Springer.

Cohen, J. (1992). A power primer. *Psychological Bulletin, 112*(1), 155.

Cortina, K. S., & Thames, M. H. (2013). Teacher education in Germany. In M. Kunter, J. Baumert, W. Blum, U. Klusmann, S. Krauss, & M. Neubrand (Eds.), *Cognitive activation in the mathematics classroom and professional competence of teachers, mathematics teacher education* (Vol. 8, pp. 49–62). New York, NY: Springer.

Gabriele, A. J., Joram, E., & Park, K. H. (2016). Elementary mathematics teachers' judgment accuracy and calibration accuracy: Do they predict students' mathematics achievement outcomes? *Learning and Instruction, 45*, 49–60.

Hoge, R. D., & Coladarci, T. (1989). Teacher-based judgments of academic achievement: A review of literature. *Review of Educational Research, 59*(3), 297–313.

Krauss, S., Blum, W., Brunner, M., Neubrand, M., Baumert, J., Kunter, M., et al. (2013). Mathematics teachers' domain-specific professional knowledge: Conceptualization and test construction in COACTIV. In M. Kunter, J. Baumert, W. Blum, U. Klusmann, S. Krauss, & M. Neubrand (Eds.), *Cognitive activation in the mathematics classroom and professional competence of teachers, mathematics teacher education* (Vol. 8, pp. 147–174). New York, NY: Springer.

Krauss, S., Brunner, M., Kunter, M., Baumert, J., Blum, W., Neubrand, M., & Jordan, A. (2008). Pedagogical content knowledge and content knowledge of secondary mathematics teachers. *Journal of Educational Psychology, 100*(3), 716–725.

Krauss, S., Lindl, A., Schilcher, A., Fricke, M., Göhring, A., Hofmann, B., Kirchhoff, P. & Mulder, R. H. (Hrsg.). (2017). *FALKO: Fachspezifische Lehrerkompetenzen. Konzeption von Professionswissenstests in den Fächern Deutsch, Englisch, Latein, Physik, Musik, Evangelische Religion und Pädagogik*. Münster: Waxmann.

Krauss, S., & Brunner, M. (2011). Schnelles Beurteilen von Schülerantworten: Ein Reaktionszeittest für Mathematiklehrer/innen. *Journal für Mathematik-Didaktik, 32*(2), 233.

Kunter, M., Baumert, J., Blum, W., Klusmann, U., Krauss, S., & Neubrand, M. (Eds.). (2013). *Cognitive activation in the mathematics classroom and professional competence of teachers. Results from the COACTIV project*. New York, NY: Springer.

Kunter, M., Klusmann, U., Baumert, J., Richter, D., Voss, T., & Hachfeld, A. (2013). Professional competence of teachers: Effects on instructional quality and student development. *Journal of Educational Psychology, 105*(3), 805–820. https://doi.org/10.1037/a0032583

Kunter, M., & Voss, T. (2013). The model of instructional quality in COACTIV: A multicriteria analysis. In M. Kunter, J. Baumert, W. Blum, U. Klusmann, S. Krauss, & M. Neubrand (Eds.), *Cognitive activation in the mathematics classroom and professional competence of teachers, mathematics teacher education* (Vol. 8, pp. 97–124). New York, NY: Springer.

Leuders, T., Dörfler, T., Leuders, J., & Philipp, K. (2018). Diagnostic competences of mathematics teachers – Unpacking a complex construct. In K. Philipp, T. Leuders, & J. Leuders (Eds.), *Diagnostic competences of mathematics teachers*. New York, NY: Springer.

McElvany, N., Schroeder, S., Hachfeld, A., Baumert, J., Richter, T., Schnotz, W., … Ullrich, M. (2009). Diagnostische Fähigkeiten von Lehrkräften: bei der Einschätzung von Schülerleistungen und Aufgabenschwierigkeiten bei Lernmedien mit instruktionalen Bildern. *Zeitschrift für pädagogische Psychologie, 23*(34), 223–235.

Prenzel, M., Baumert, J., Blum, W., Lehmann, R., Leutner, D., Neubrand, M., et al. (Eds.). (2004). *PISA 2003. Der Bildungsstand der Jugendlichen in Deutschland—Ergebnisse des zweiten internationalen Vergleichs*. Münster, Germany: Waxmann.

Schrader, F. W. (1989). *Diagnostische Kompetenzen von Lehrern und ihre Bedeutung für die Gestaltung und Effektivität des Unterrichts*. Frankfurt am Main, Germany: Lang.

Spinath, B. (2005). Akkuratheit der Einschätzung von Schülermerkmalen durch Lehrer und das Konstrukt der diagnostischen Kompetenz: Accuracy of teacher judgments on student characteristics and the construct of diagnostic competence. *Zeitschrift für pädagogische Psychologie, 19*(1/2), 85–95.

Südkamp, A., & Praetorius, A. K. (Eds.). (2017). *Diagnostische Kompetenz von Lehrkräften: Theoretische und methodische Weiterentwicklungen*. Waxmann Verlag.

Competences of Mathematics Teachers in Diagnosing Teaching Situations and Offering Feedback to Students: Specificity, Consistency and Reification of Pedagogical and Mathematical Discourses

Irene Biza, Elena Nardi, and Theodossios Zachariades

In the study we report in this chapter, we investigate the competences of mathematics pre- and in-service teachers in diagnosing situations pertaining to mathematics teaching and in offering feedback to the students at the heart of said situations. To this aim, we deploy a research design that involves engaging teachers with situation-specific tasks in which we invite participants to: solve a mathematical problem; examine a (fictional yet research-informed) solution proposed by a student in class and a (fictional yet research-informed) teacher response to the student; and, describe the approach they themselves would adopt in this classroom situation. Participants were 23 mathematics graduates enrolled in a post-graduate mathematics education programme, many already in-service teachers. They responded to a task that involved debating the identification of a tangent line at an inflection point of a cubic function through resorting to the formal definition of tangency or the function graph. Analysis of their written responses to the task revealed a great variation in the participants' diagnosing and addressing of teaching issues – in this case involving the role of visualisation in mathematical reasoning. We describe this variation in terms of a typology of four interrelated characteristics that emerged from the data analysis: *consistency* between stated beliefs/knowledge and intended practice, *specificity* of the response to the given classroom situation, *reification of pedagogical discourses,* and *reification of mathematical discourses.* We propose that deploying the theoretical construct of these characteristics in tandem with our situation-specific task design can contribute towards the identification – as well as reflection upon and development – of mathematics teachers' diagnostic competences in teacher education and professional development programmes.

I. Biza (✉) • E. Nardi
University of East Anglia, Norwich, UK
e-mail: I.Biza@uea.ac.uk

T. Zachariades
National and Kapodistrian University of Athens, Athens, Greece

T. Leuders et al. (eds.), *Diagnostic Competence of Mathematics Teachers*,
Mathematics Teacher Education 11, DOI 10.1007/978-3-319-66327-2_3

1 Introduction

Mathematics teachers (pre- and in-service) typically engage with teacher education programmes towards their preparation for the classroom and/or the enhancement of their subject and pedagogical knowledge. With the study we present in this chapter – parts of which we have reported also in Biza, Nardi and Zachariades (2007, 2009), Nardi, Biza and Zachariades (2012), and Zachariades, Nardi and Biza (2013) – we aim to contribute to insights into teachers' benefits from these programmes through the investigation of their competences in diagnosing issues pertaining to mathematics teaching situations and in addressing these issues in the feedback they state they would offer to students.

Research has reported the overt discrepancy between theoretically and out of context expressed teacher beliefs about mathematics and pedagogy and actual practice (e.g. Speer, 2005; Thompson, 1992). Speer (2005) claims, for example, that instead of discussing beliefs and teaching practices in the abstract, a discussion of these in a concrete context can provide shared understanding between researchers and participating teachers of the beliefs that are attributed by researchers to teachers. With this observation in mind, our study makes the case for a situation-specific task design that explores teachers' beliefs, knowledge and competencies, and the relation of these to practice, through teacher engagement with, and reflection on, fictional yet realistic and research-informed teaching situations (Biza et al., 2007). In the situation-specific task we discuss in this chapter, we invite participants to:

(a) Solve a mathematical problem
(b) Examine a solution proposed by a student in class
(c) Examine a teacher response to the student
(d) Describe the approach they themselves would adopt in this classroom situation

Specifically, this task (*Tangent Task – N, as in Tangent Task-New version,* Fig. 1) is a variation of the *Tangent Task* we have used in our previous research (Biza, et al., 2009; Nardi et al., 2012) and describes a teaching situation in which a teacher and a student discuss whether a line is a tangent line of a function graph.

The introduction and the development of these tasks started in 2005. Each task is based on a teaching situation, which is fictional, yet derived from findings in prior research (in the case of the task in this chapter, e.g., Biza, Christou, & Zachariades, 2008). Over the years, we have deployed various versions of the situation-specific task design proposed in this chapter. We firstly introduced this type of task in Biza et al. (2007). That version involved a classroom situation and a single student response. In Nardi et al. (2012) we reported analyses from the use of a modified version that involved several student responses. The version we present in this chapter (Tangent Task-N) involves also a teacher reaction to student responses as we explain in detail in Sect. 3. All three versions outlined above are geared deliberately towards an examination of primarily mathematical issues. A fourth version (Biza, Nardi, & Joel, 2015) embroiders also classroom management and affective issues.

In a class of Year 12 students specialising in mathematics, the teacher gave to the students the following problem:

"Examine whether the line with equation $y = 2$ is tangent to the graph of function f, where $f(x) = 3x^3 + 2$"

A student responded as follows:

"I will find the common points between the line and the graph solving the system:

$$\begin{cases} y = 3x^3 + 2 \\ y = 2 \end{cases} \Leftrightarrow \begin{cases} 3x^3 + 2 = 2 \\ y = 2 \end{cases} \Leftrightarrow \begin{cases} 3x^3 = 0 \\ y = 2 \end{cases} \Leftrightarrow \begin{cases} x = 0 \\ y = 2 \end{cases}$$

The common point is $A(0, 2)$. The line is tangent of the graph at the point A because they have only one common point (which is A)."

The following dialogue then took place in the classroom:

T (Teacher): "The parabola $y = x^2$ and the line $x = 0$ have only one common point, the point (0, 0). Is the line $x = 0$ tangent of the parabola at this point?"

The student sketches the parabola and the line on the board and answers:

S (Student): "No, it isn't, because the line cuts the parabola at this point."

T: "OK. In our case (the teacher shows the problem in question) what is happening?"

The student sketches the following graph and answers:

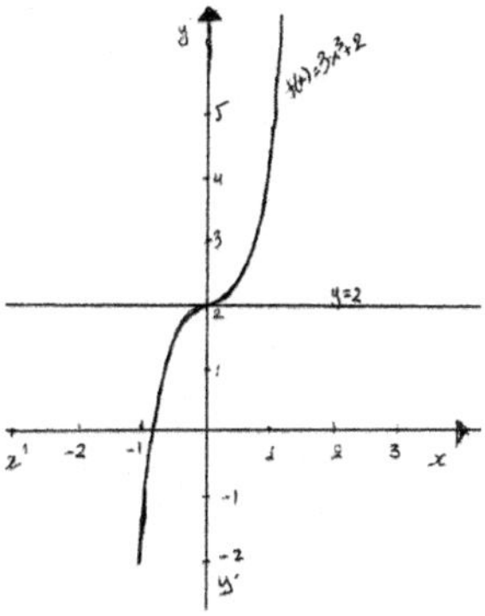

S: "As we see from the graph, the line $y = 2$ cuts the curve $y = 3x^3 + 2$ at the point (0, 2). So, the line is not a tangent of this curve."

T: "This is correct but you also need to justify it algebraically. Even if a graphical understanding of functions is particularly useful, you should not forget that it is not always possible to use graphical representations and that you should learn to solve problems also algebraically."

Questions

1. How do you evaluate the teacher's management from
   a) a mathematical perspective?
   b) a didactical perspective?
   c) an epistemological perspective, especially regarding the teacher's beliefs about the role of visualization in mathematics?
2. If you were the teacher, how would you manage the situation following the student's answer to the problem?

Justify your responses.

Fig. 1 The *Tangent Task-N* used in the study

A fifth version (Nardi, Healy, Biza, & Fernandes, 2016) presents actual data as written and video excerpts (and, in the case of the aforementioned study with Healy and Fernandes, the focus is on inclusion in mathematics lessons of students with disabilities). Finally, at the time of writing, and in resonance with emerging work in this area (e.g. the lesson plays in (Zazkis, Sinclair, & Liljedahl, 2013)), we have started to explore the potency of teachers' selecting and composing their own tasks.

We invited 23 mathematics graduates enrolled in a post-graduate mathematics education programme to engage with this task. Analysis of their written responses to the task revealed a great variation in the participants' competences in diagnosing teaching issues – in this case involving the role of visualisation in mathematical reasoning and in addressing these issues in the feedback they would offer to their students. We describe this variation in terms of a typology of four interrelated characteristics. Here we illustrate these characteristics through examples from our data analysis. We also aim to illustrate how the use of this typology can offer insights into the relationship between the participating teachers' stated beliefs/knowledge and their intended action in the classroom.

In the following sections, first, we consider the role of situation-specific task design in teacher education and research. Then, we discuss the task design in our study and Tangent Task-N (Fig. 1). Also, we present the context of the study, its methodology and the participants. The presentation of the results of the study follows, by offering a typology of four interrelated characteristics that has emerged from the data analysis: consistency between stated beliefs/knowledge and intended practice, specificity of the response to the given classroom situation, reification of pedagogical discourse, and reification of mathematical discourse. We conclude with a discussion of how our situation-specific task design and this typology of four interrelated characteristics can contribute towards the identification – as well as reflection upon and development – of mathematics teachers' diagnostic competencies. We especially make the case for the potency of this task design / typology combination for teacher education and professional development programmes – and we embed our proposition in current developments in this area.

2 Situation-Specific Tasks in Teacher Education and Research

> …the fundamental issue in working with teachers is to resonate with their experience so that they can imagine themselves 'doing something' in their own situation (Watson & Mason, 2007, p. 208)

The design and use of tasks for pedagogic purposes is at the core of mathematics teacher education and mathematics education research (Artigue & Perrin-Glorian, 1991; Sierpinska, 2003). Especially, in the field of mathematics teacher education, significant attention has been paid to the nature, role and use of tasks. Recent work, such as parts of the *Handbook of Mathematics Teacher Education* (Tirosh & Wood,

2009) has focused on integrating tasks into the processes of teacher education. Also, a special issue of *Journal of Mathematics Teacher Education* (2007) edited by Zaslavsky, Watson and Mason and the book edited by Zaslavsky and Sullivan (2011), signal this interest.

In the literature, the word task is used in different ways (Christiansen & Walter, 1986; Leont'ev, 1975; Mason & Johnston-Wilder, 2006) and often conveys that tasks are mediating tools for teaching and learning mathematics. In the case of teacher education, a task can be used to trigger teachers' reflection and to explore their mathematical knowledge for teaching as well as their pedagogical and epistemological perceptions and beliefs. An appropriately designed task, which addresses complex purposes, affords opportunity to engage with aspects of mathematics, didactical strategies, pedagogical theory and epistemological beliefs. We see all these aspects as crucial in teachers' diagnostic proficiency when they deal with unexpected situations in the classroom that demand immediate reaction.

A substantial body of work in mathematics education explores the use of *cases*, that is, "any description of an episode or incident that can be connected to the knowledge base for teaching" (Carter, 1999, p. 174), in mathematics teacher education and research (see a comprehensive review in Markovits & Smith, 2008). Shulman (1992, p. 28) envisioned "... case method as a strategy for overcoming many of the most serious deficiencies in the education of teachers. Because they are contextual, local, and situated – as are all narratives – cases integrate what otherwise remains separated."

Over the years, this key idea has gained substantial momentum in mathematics teacher education, whether in the shape of brief classroom situations used as prompts (e.g. Erens & Eichler, 2013; Dreher, Nowinska, & Kuntze, 2013) or, in the shape of more extended "imagined" classroom dialogues, such as Zazkis et al.'s (2013) "lesson plays". As Zazkis et al. write "[w]ith this imagination, attention and awareness are developed in 'slow motion', having a complete control of the situation and ability to replay or redress it, rather than 'thinking on one's feet' and making in the moment decisions" (p. 29). The task design we put forward in our study resonates well with these works.

3 Situation-Specific Task Design and the *Tangent Task-N*

The type of task we use in the study presented in this chapter is a development of former task designs, which were studied in Biza et al. (2007, 2009) and Nardi et al. (2012). We use tasks of these types for research and teaching purposes, including formative and summative assessment, in under/postgraduate programmes in mathematics education, particularly those run by the third author. Key to our task design and use is to provide pre- and in-service mathematics teachers with opportunities to engage with plausible, teaching situations from the secondary mathematics classroom in order to further our understanding of – and strengthen our ways of influencing – these teachers' knowledge and beliefs.

As we exemplify with the *Tangent Task-N* in Fig. 1, a situation-specific task in our studies describes a teaching situation that is triggered by a mathematical problem given to students. The situation includes a response from one or more students and in some versions a dialogue between students and teacher or/and a reaction from a teacher. On the basis of this teaching situation we create a list of questions on which we invite teachers to reflect on potential issues and how they would react in a similar situation.

Our tasks are grounded in learning and teaching issues that previous research and experience have highlighted as seminal. Although our tasks are addressing teacher education needs, in their design we also consider student learners and their needs that teachers should be prepared to address in their actual teaching practice. For example, at the heart of the teaching situations in our tasks are pivotal moments in the growth of learners' mathematical thinking. These moments are akin to what Leatham, Peterson, Stockero and Van Zoest (2015) call *Mathematically Significant Pedagogical Opportunities to build on Student Thinking* (MOSTs), which are "instances of student thinking that have considerable potential at a given moment to become the object of rich discussion about important mathematical ideas" (p. 90). Specifically, we see as a core aim of our work to identify and facilitate the ways in which teachers recognise MOSTs and optimise these opportunities as they diagnose the issues in a classroom situation and transform their beliefs and knowledge (mathematical and pedagogical) into practice. In this sense, the situations in our tasks satisfy the three characteristics of MOSTs: "student mathematical thinking, mathematically significant, and pedagogical opportunity" (p. 91).

We suggest the use of these tasks in teacher education to explore, assess and develop teachers' *Mathematical Knowledge for Teaching* (*MKT*, Hill & Ball, 2004), especially in relation to their diagnostic ability and intended practice when confronted with realistic teaching situations. Additionally, with these tasks we aim to address the complex set of considerations that teachers take into account when they determine their actions. To this aim, we draw on what Herbst and colleagues (e.g. Herbst & Chazan, 2003) describe as the *practical rationality of teaching (PRT)*. We delve into these considerations and in our previous research we identified a *spectrum of warrants* (*SW*) secondary mathematics teachers put forward in order to justify the decisions they intend to make in their classroom: *empirical–personal, empirical–professional, institutional–curricular, institutional–epistemological, a priori–epistemological, a priori–pedagogical* and *evaluative* (Nardi et al., 2012).

Here, we focus on elaborating further the interaction of teachers' expressed knowledge and beliefs with what they diagnose as an issue in a teaching situation and how they intend to address this issue. We are interested in teachers' competences in the identification of both mathematical and pedagogical issues. To this aim, we draw on what Rowland and colleagues (Turner & Rowland, 2011) describe as *Foundation* – one of the four features of the *Knowledge Quartet* (*KQ*), with the other three being *Connection*, *Transformation* and *Contingency* – namely, amongst others, the "overt subject knowledge, theoretical underpinning of pedagogy, use of terminology" (p. 200). Additionally, we see Ball and colleagues' (Ball, Thames, & Phelps, 2008) *Horizon Knowledge* (*HK*) – "an awareness of how mathematical topics are related over the span of mathematics included in the curriculum" and "the vision useful in seeing connections to much later mathematical ideas" (p. 403) – as a useful component of mathematical knowledge for teaching that brings together mathematical and curricular content.

With the design of the task we focus on in this chapter, we aim to explore whether the teacher can diagnose a student's mathematical error, what their pedagogical intentions are, and how they evaluate the pedagogical approach followed by another teacher. As we elaborated earlier, the student response to the mathematical problem and the teacher's response to the student are grounded in issues identified as seminal in previous research. Through engagement with the tasks we aim to explore various aspects of teachers' knowledge (mathematical, pedagogical and epistemological) related to their diagnostic competences and their ability to support their views and choices, especially when juxtaposed to those of another teacher. In this respect, in designing these tasks we bear in mind the following:

- The mathematical content of the task concerns a topic or an issue that is known for its subtlety or for causing difficulty to students (from literature and/or previous experience) (*MOSTs: student mathematical thinking, mathematically significant*)
- The student's response reflects this subtlety (or lack of) or difficulty and provides an opportunity for the teacher to reflect on and demonstrate the ways in which s/he would help the student achieve subtlety or overcome difficulty (*MOSTs: pedagogical opportunity*)
- The teacher's pedagogical approach concerns mathematical, pedagogical and epistemological issues that are known for their subtlety or for being challenging to teachers (*PRT, SW*)
- Mathematical content and student/teacher responses provide a context in which teachers' knowledge, beliefs and intended practices (mathematical, pedagogical and epistemological) are allowed to surface (*MKT, HK, KQ*)

The mathematical content of the *Tangent Task-N* presented in Fig. 1, draws on two issues identified in research (e.g., Biza et al., 2008; Castela, 1995) in relation to student learning of tangent line: (a) students often believe that one common point between a line and a curve is a necessary and sufficient condition for tangency; and, (b) students often see a tangent as a line that keeps the entire curve in the same semi-plane (MOSTs: *student mathematical thinking, mathematically significant, pedagogical opportunity*). The teacher's response in a dialogic format in this situation draws on results of preceding studies that identified teachers' views on the role of visualisation and their perceptions about the tangent at an inflection point of a curve (e.g. Biza et al., 2009) as well as analysis of their warrants in arguing for or against certain pedagogical approaches (Nardi et al., 2012). We consider that this situation can offer the opportunity to discuss mathematical, pedagogical and epistemological issues in the teacher response (*PRT, SW*) and provide a context in which participants' knowledge, beliefs and intended practices (mathematical, pedagogical and epistemological) are allowed to surface (*MKT, HK, KQ*).

In addition, our aim is to explore the participants' knowledge and beliefs, not only through the way in which they would diagnose the issues and tackle the teaching situation, but also in relation to the arguments they use in evaluating another teacher's approach. Comparing the participants' responses to the two questions (Question 1: comment on the teacher's response; Question 2: describe how *you* would tackle the situation), we can explore and identify also possible discrepancies between their stated beliefs/knowledge (as evident mainly but not exclusively, in the critique of the teacher's approach in the first question of the task), and their intended feedback to the student (as evident, also mainly but not exclusively, in their response to the second question).

4 Participants, Data Collection and Data Analysis

Participants were 23 mathematics graduates enrolled in a post-graduate mathematics education programme, many already in-service teachers. The participants attended a mathematics education module with a focus on the teaching of Calculus taught by the third author, as part of their studies within this post-graduate programme in the Department of Mathematics of a Greek University and they had engaged with tasks of this type during the module. They responded to this task during the module's written examination and, in a note attached to the exam documents, all agreed with the use of their responses for research purposes.

We recognise that, in the course of their engagement with the task, the participants were not in the classroom and they had some time to think about their reaction. However, we consider that, exactly because the participants were not in the classroom and did not respond under its pressing circumstances, their responses may be more reflective. We acknowledge that this distancing from the classroom may render their responses slightly wishful. We also note that the responses can offer substantial evidence of the insight into the teaching and learning of mathematics that the participants gained during their engagement with the module. In this respect, the responses can also be seen as reflecting the respondents' keenness to showcase how much they have learned in order to achieve the highest possible mark in the examination. Finally, we also acknowledge the potential tension between assessing participants' performance in the task – as the third author is expected to do towards meeting the module assessment requirements – and examining their beliefs, knowledge and competences in a non-deficit and non-judgemental way for the purposes of our research. Our awareness of these issues has implied that we are conducting the analyses of these scripts with caution and moderation. This caution is evident also in the several rounds of triangulating our analyses within the team (and we also note that the first and second authors have no involvement with the module).

In the spirit of a data-grounded approach (Charmaz, 2000) to the analysis of the scripts, our initial scrutiny of the 23 responses to the *Tangent Task-N* led to preliminary observations concerning the ways in which the participants diagnose issues in a teaching situation and address these issues in their intended teaching practice. Analysis of their written responses to the task revealed a great variation in this diagnosing and addressing of teaching issues – in this case involving the role of visualisation in mathematical reasoning – and led to the observation of potential discrepancies between participants' stated beliefs and intended practice (Zachariades et al., 2013). The elaborate analysis that followed resulted in a typology of four characteristics of participants' responses. In what follows, we outline the preliminary observations that led to the emergence of these characteristics. We then illustrate these characteristics with examples from the data.

5 Results: Emergence of Four Characteristics

From the preliminary analysis of the participants' written responses, evidence emerged about their diagnostic competencies, especially in relation to the meanings they attribute to tangency. Also, evidence emerged in relation to their pedagogical intentions in the teaching about tangency and their epistemological perceptions about the role of visualisation. Also, through their evaluation of the teacher's response (question 1), some of their pedagogical beliefs emerged. The participants' intended practices in relation to the feedback they would offer to students in a similar situation emerged mainly from their responses to question 2. We note that during the post-graduate module that these participants attended, the role of visualisation in mathematics, and especially in teaching Calculus, had been discussed extensively. Also, most of the mathematics education terminology and the mathematical elements participants used in their responses had been introduced across the modules of the post-graduate mathematics education programme they were enrolled.

In our preliminary analysis, the common belief of the participants, as it emerged from their written responses to question 1, was that visualisation plays a very important role in the teaching of Calculus, because of its potential to support student meaning making. In consistency with their response to question 1, most participants wrote in their response to question 2 that they would use graphs to support student meaning making of tangency. Many of them sketched some graphs in their scripts. Most scripts therefore demonstrated consistency with regard to appreciation and use of visualisation. Some scripts, however, were not as internally consistent as we discuss in more detail in the following section. This initial observation of potential discrepancies between stated beliefs and intended practice (Zachariades et al., 2013) became the focus of our data analysis and we started orientating our efforts towards tracing consistency and coherence within each script.

The elaborate analysis of each one of the scripts that followed led to certain observations in relation to: participants' reflection on the interface of mathematical, pedagogical and epistemological issues; the use of the mathematics education terminology; and, their engagement with the mathematical elements of the problem. At this juncture, we also started to differentiate between pedagogical and didactical intentions in the scripts,[1] with the latter designating practices specific to the classroom situation or mathematical topic under scrutiny and the former designating the respondent's broader pedagogical practices. Another observation in this phase of our analyses was that the mathematical competency of some of the scripts (e.g. ones with a plethora of examples about tangency) was in itself of limited effect when the pedagogical purpose of this plethora was not clear.

In sum, this more elaborated analysis of the scripts suggested four characteristics that can act as theoretical lenses through which to examine the scripts:

- *Consistency*: how *consistent* a response is in the way it conveys the link between the respondent's stated beliefs and their intended practice,
- *Specificity*: how *contextualised and specific* a response is to the teaching situation in the task,
- *Reification of pedagogical discourse*: how *reified*[2] the *pedagogical discourse* of the response is in order to describe the pedagogical and didactical issues of the classroom situations and the intended practice presented in the script, and
- *Reification of mathematical discourse*: how *reified* the *mathematical discourse* of the response is in relation to the identification of the underpinning mathematical content of the classroom situations and the transformation of this mathematical content into the intended practice presented in the script.

In the section that follows we elaborate each characteristic with illustrative examples from our data analysis.

6 Illustration of the Four Characteristics Through the Data

To illustrate the four characteristics we now present evidence from the 23 scripts. Where necessary, responses have been translated verbatim from Greek. The two main criteria for selecting these excerpts were typicality across the 23 scripts and clarity of illustration.

[1] The original version of the task is in Greek. The term didactical in the context of this task, and more broadly in the context of the post-graduate programme attended by the participants, is used to denote pedagogical strategies related to specific mathematical topics (as in, for example, didactics of Calculus). In the programme the term was also used with the sense that it has within the Theory of Didactic Situations (Brousseau, 1997), for example, as in didactic contract.

[2] Our use of the term "reification" takes cue from discursive perspectives such as Sfard's (2008) where reification is defined as the gradual turning of processes into objects. Discourses, Sfard writes, change in a "chain of intermittent expansion and compression" (p. 118). Reification is the key element of compression which can be endogenous – resulting from saming within one particular discourse – and exogenous which "conflates several discourses into one" (p. 122). Reification is a response to what discursive researchers see as our innate "need for closure" (p. 184) in our use of signifiers and brings at least two potent gains: increasing the communicative effectiveness of discourse and increasing the practical effectiveness of discourse. In our analyses, we are interested particularly in the extent, and ways, in which participants' discourse (for example, as evident in their use of mathematics education terminology) is reified.

6.1 Consistency

We analysed the scripts in relation to participants' stated beliefs as these were expressed throughout their response and especially through their evaluation of the teacher's reaction (question 1). Also, we analysed participants' intended practices in relation to the feedback they would offer to students in a similar situation. This emerged mainly from their responses to question 2. In terms of the relationship between stated beliefs and intended practice, we saw the responses as being: *consistent* in the way they related stated beliefs and intended practice, or with *elements of consistency but incomplete actualisation* of the stated beliefs in the intended practice, or *inconsistent.*

One example of a response that suggested consistency in relation to the role of visualisation was participant [2]'s. In her response to question 1c, she agrees with the views of the teacher of the task (thereafter Teacher) regarding the role of visualisation. Her interpretation of these views is:

> It [visualisation] is useful – as [the Teacher] says – to understand functions through their graphs, however we should not restrict ourselves only to them. There are times when they [graphs] cannot help us. The algebraic representation of the problem should always follow as it is the most accurate and rigorous response. We use visualisation as a tool and actually a very-very good [tool] but this does not suffice as a proof method in mathematics.

In her response to question 2, she is consistent with the above views: she would follow similar actions to the ones followed by the Teacher but, in the end, she would explain to the student that the line $y = 2$ is the tangent of the graph by "applying the definition which can be found on page 212 of the textbook" (the respondents were allowed access to curricular materials during their engagement with the task). She solves the problem with the calculation of derivative and she concludes that:

> Then we define as tangent of C_f at the point A (0, 2), the line ε which goes through the point A (0, 2) and has slope $\lambda = 0$. This line is: $y = 2$.

Later she suggests more examples of functions in her aim to "help the students escape from the 'initial image' of tangent line they have in their mind which is the <u>circle tangent</u>" [her underlining].

The same participant ([2]) appears less consistent in terms of her stated pedagogical preferences in her critique of the Teacher in question 2b. There she agrees with the Teacher's approach and she appreciates certain elements of *student*-centeredness in the Teacher's approach: dialogue invitation to the student; enquiry-based approach; and, encouragement to the student to find the solution independently (the extended excerpt only summarised here). However, later in her response to question 2, her discourse is distinctly *teacher*-centred: although she states that she would follow a similar approach to that of the Teacher, she then writes that *she* "would explain to the student" that the line is a tangent by applying the definition; *she* "would highlight" that a tangent line can cut the graph of a function; *she* "would give more examples"; and, *she* "would highlight" that the tangent can have more than one common point with the curve (the emphasis on "she" is ours). In sum, although she mentions that she is interested in helping her students escape from the constraining "initial image" of tangent line, she does not mention any pedagogical approach on how she would do so through engaging students in a student-centred way (her approach of choice according to her earlier statement).

Several other scripts suggested inconsistency between stated beliefs and intended practice. Participant [18], again regarding the use of visualisation, seemed to appreciate the use of graphs in his critique to the Teacher but without implementing this appreciation in his suggested approach. In his response to question 1c, he writes:

> Visualisation is a very important part of mathematics because through this, intuitions, conjectures and concept images form and give to the student the possibility to understand concepts better.

However, in his response to question 2, he does not use or refer to any visualisation. He uses only formal mathematics and completes the algebraic solution initiated by the student. So, in the above excerpt, the participant expresses a view about visualisation which echoes what was discussed, and perhaps prevailed, during the module. His intended practice though is completely different.

Other participants seem to appreciate formal and graphical approaches in their critique to the Teacher but without implementing such appreciation in their suggested approach. For example, participant [16] expresses her appreciation for the contribution that diverse modes of thinking can make in mathematics meaning making:

> The connection between the embodied and proceptual mode of thinking is necessary and in order to materialise it [the connection] student practice with graphical representation of mathematical objects and with the transition from these [graphical representations] to the formal [representations] and vice-versa is required. The visual representations help in the understanding of mathematical concepts, in forming conjectures, in describing a mathematical result [...]

Later, she adds:

> However the mathematical truth is revealed only through formal proof! So, it is necessary for the student to learn how to make the transition between visual representations and formal proofs.

In the above, we can see participant [16]'s appreciation for the role of visual representations in "understanding mathematical concepts, forming conjectures and describing a mathematical result". But, also, we can discern her requirement for formal proof towards the securing of mathematical "truth". Also, she highlights the importance of the "transition" (notably mentioned in two places in the script) between the visual and the formal modes of mathematical thinking.

In the light of this evidence we would expect a similar approach in her response to question 2. However, her response to question 2 includes a suggestion of graphs similar to those presented in the task, references to the tangent line as the limiting position of secants, comparison of the tangent line in Euclidean and Analytic Geometry with this in Analysis and, finally, a suggestion of more examples of graphs and tangent lines. In the entire response there is no reference to the formal definition of the tangent line or to the transition from the visual to the formal modes. We can see in her response the effort to materialise her appreciation of the visualisation and the didactical opportunities this can offer. We cannot see, however, complete actualisation in her intended practice of her earlier stated beliefs about the necessity "for the student to learn how to make the transition between visual representations and formal proofs".

We identified elements of inconsistency in about half of the scripts and it is the study of the different types of (in)consistency that led to further elaboration and identification of the three characteristics that we discuss in what follows.

6.2 Specificity

We found a great variation in the ways in which the responses ranged from being highly specific to the classroom incident in the task to being only peripherally related to the incident.

An example of a highly specific script is participant [13]'s. His approach is well structured and with a clear focus on the specific mathematical problem and the mathematical issues related to this problem. His response is in Table 1, chunked (by us) in 13 utterances (left-hand side column of Table 1). In the right-hand side column, we cite scans of the accompanying images from his script.

In his response, [13] first suggests a similar approach to that of the Teacher (1). Then (2–9), because he wants to reconstruct the student's perception that the tangent cannot cut the function graph, he suggests an approach through example construction that starts from the circle and leads to a graph similar to the function of the task. Then (10–12), with more examples, he challenges the perception of the "one common point" as a necessary and sufficient condition for tangency. Finally (13), he asks the student to deal again with the initial problem by using analytical methods. We characterised [13]'s response as highly specific to the incident in the task. We highlight the following as warrants for this characterisation:

- He aims to facilitate the transition from the geometrical tangent (circle tangent) to analytical tangent (tangent to a graph). He suggests a series of steps through which the circle tangent is transformed to a tangent to an inflection point in an intuitive and natural way.
- He discusses clearly both the issues of "one common point" and "the tangent can cut the graph" with a series of examples connected to the mathematical problem and the dialogue in the task.
- He constructs his response to question 2 (2–9) in what appears to be an imagined dialogue with the student. Although, this dialogue is not fully developed – some lines are ambiguous ("maybe", (7)) or incomplete ("NOW?" (9)) – we see these utterances as an indication of his effort to tailor his reaction specifically to the student in the task.

Table 1 Participant [13]'s response to question 2

1	Until the point [where the teacher says "this is correct"] except the "this is correct" I would have done the same but for different reasons. The student gave a correct response [when he/she said] tangent with wrong reasoning and I would have given the examples to correct this * (*not because I believe that [the line] is not tangent as the teacher [did])	
2	Now I make the following figures	
3	Two circles 1 common point [pointing to]	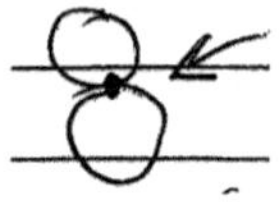
4	I sketch the tangent there [pointing to]	
5	Is it tangent? Yes	
6	I erase a bit [pointing to]	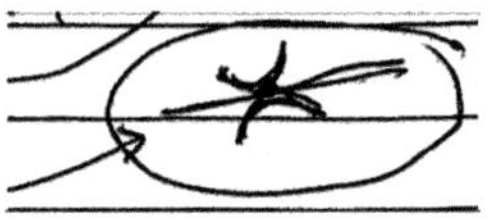
7	Is it tangent? Yes (maybe)	
8	I erase a bit more [pointing to]	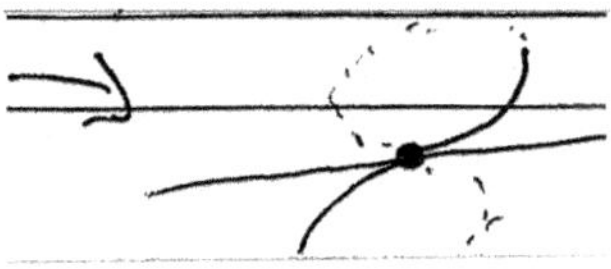
9	Now?	
10	Then I would ask what [his/her] opinion is regarding the criterion of "one common point" eventually since [he/she] saw that it does not work in parabola. [his emphasis]	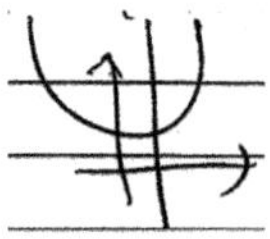
11	[He/she] should consciously understand that it [the "one common point"] is not a criterion and it is neither sufficient nor necessary condition namely: It [the line] can have 1 common point without being a tangent or it [the line] has many points in common and be a tangent x^3	
12	Also the example x^2, $x \geq 0$ and 0, $x < 0$ [the tangent] at zero [has] infinite points in common [with the graph]	
13	Finally I would ask [the student] to write the solution with analysis [by using an analytical method]	

- After dealing with a series of examples, mainly in a visual and intuitive way, he outlines his plan to request the student to revisit the task by applying analytical methods. This is in consistency with what he mentioned in his response to question 1c:

> [Visualisation] gives us the first steps to gain an image but afterwards everything should be proved analytically. Of course, I believe that the role of visualisation is very essential and in my teaching I would like to start with this [visualisation] and then step by step to TRANSLATE into algebra in order to show to the students how to use the figures [his capitalisation].

We note that most participants mentioned the transition from the geometrical tangent (circle tangent) to analytical tangent (tangent to a graph) in their response to question 2. Some suggested this transition in a theoretical way without indicating how they would materialise it (see the following example, from [10]'s response) and in some responses the proposed examples appeared disconnected to the rest of the response or irrelevant.

In contrast to participant [13]'s highly systematic approach, the following response from participant [10] does not address the specific mathematical problem and the mathematical issues related to this problem:

> The student has basic *misconceptions* regarding the concept of tangent line under the influence of the circle tangent and regards that the line $y = 2$ and the function graph of f since they have one common point then $y = 2$ would be a tangent of f. Thus, *I would focus on this point* and after *discussion* with the student and by offering [to the student] certain *examples*, [in which examples] what [student] says is not valid, I would try to make [the student] to understand his mistake on his own. Additionally, an *expansion* of what he [the student] knows about the concept of tangent should be done to all function graphs. The *counterexamples* I would suggest to him [the student] would have had a specific aim, namely to make the *expansion* of what he knows about the tangent of circle to all the function graphs. Thus, through the *resolving of his misconceptions* and with the help of function graphs of the function f [the student] would have been driven eventually to the solution of the initial problem. My role would have been to *guide and help* throughout *without giving the responses* to the student directly. The aim is through the discussion that [the student] will find the solution. [our italics]

We note the following in participant [10]'s response:

- She addresses student's potential difficulties with tangent line and the influence of the circle tangent.
- She has commendable didactical intentions *in relation to the tangent line*:
  - She identifies what she sees as the student's "misconception".
  - Her approach focuses on the "misconception", especially in relation to prior knowledge (circle tangent).
  - She aims to use examples and counterexamples.
  - She aims to lead the student towards an expansion of what he/she knows from the circle tangent.
- She also has commendable *overall* pedagogical intentions:
  - She aims to trigger discussion with the student.
  - She aims to let the student discover the solution on their own but with some guidance from her.
- However, there is no concrete evidence of *how* she would materialise these very commendable intentions.

One determinant of whether scripts achieved the *consistency* and *specificity* we have examined in this and the previous section (6.1) was the extent, and ways, in which participants' discourse appears *reified*. We examine said *reification* (first *of pedagogical discourse*, then *of mathematical discourse*) in what follows.

6.3 Reification of Pedagogical Discourse

The influence of the pedagogical discourse (e.g. mathematics education terminology) that the respondents had become familiar with during the module, and the post-graduate programme more broadly, is discernible in the scripts in, for example, the explicit use of terms such as *constructivism* or *conceptual change*. We found a great variation in the reification of this discourse in such use: in some responses such use seemed *essential* in diagnosing student needs and in shaping intended practice and in some such use was almost *superfluous* or even *inaccurate*.

Participant [6]'s response is a typical example of use that we saw as *essential*. He is using a distinctly constructivist language in his evaluation of the Teacher's management in the task. For example, in his response to question 1c he mentions:

> It seems that the teacher believes that graphs and representations in general support the understanding of mathematical ideas. From the perspective of *construction* of mathematical thinking this position is correct [our italics]

Later on, in his response to question 2, he discusses the examples of the task and he adds:

> [...] with other examples I would help [the student] to materialise the *conceptual change* of the tangent concept. The aim is to enrich the *concept image* of tangent the student has in mind. At the moment the student has a very poor *concept image*, [this concept image] is restricted to the circle tangent and the tangent of parabola and ellipse. [our italics]

Further on, he also adds that:

> In general I would *trigger a discussion* with the whole class regarding the tangent, because the *gradual filling* of the *concept image* is a *longitudinal process* and it needs many *examples* [our italics]

Again there is evidence here of: the appreciation for exemplification and its role in student understanding; acknowledgement that concept image formation is a longitudinal process; and, the importance of classroom discussion.

In the above excerpts, [6]'s response is inextricably linked to several mathematics education theoretical constructs that he became familiar with during the programme. Here are some of the key references that were used in the programme in relation to these constructs:

- the role of prior knowledge and experiences in this prior knowledge, especially students' understanding of tangent line: students' "concept images" for tangent line (Biza et al., 2008; Vinner, 1991); the influence of prior knowledge (circle, conic sections) (Biza et al., 2008);

- carrying out "the conceptual change" (Vosniadou & Verschaffel, 2004) with the help of the teacher;
- example use (Watson & Mason, 2005);
- students' "concept images" (Tall & Vinner, 1981).

Overall, we see evidence of reified mathematics education discourse in participant [6]'s response: he uses relevant terms and does so accurately; and, his familiarity with the respective mathematics education theories seems to shape his discourse about his intended practice (even though we also note that he does not suggest any further and more concrete ways to demonstrate how he would materialise his stated intentions).

Evidence of less *reified pedagogical discourse* in relation to the use of mathematics education terminology can be seen in those responses where such use seems *superfluous*, namely it comes across as paying little more than lip-service to the course contents. Even more, in a small number of responses, use of mathematics education terminology comes across as plainly *inaccurate*. For example, participant [8] writes in her response to question 1b:

> From the didactical perspective T [teacher] aims to apply the approach through the *epistemological obstacles of Brousseau*. We start from the previous knowledge (one common point C_f with tangent) and with an appropriate *epistemological obstacle* the need of the new knowledge would emerge. [our italics]

During the post-graduate programme, Brousseau's work had been discussed mostly in terms of the concept of *didactic contract* (1997), and the construct of *epistemological obstacles* mostly through the work of Sierpinska (1987). Apart from this misattribution, *epistemological obstacles* seem to be treated by participant [8] as synonymous to *cognitive conflicts*. This is also evident in the way that this participant describes the examples that the Teacher proposes in the task as "obstacles" ("...the T. [teacher] needs to find a better obstacle and not return to the algebraic formalism", she writes). We see the excerpt from her response to question 2 also as further evidence of the misnaming reference to *epistemological obstacles*:

> If I had this student's solution I would be very happy because as I said above I could use the didactical [approach] through the *obstacles* of *Brousseau*. The student triggers me to offer an obstacle like the T. [teacher] in the question. [our italics]

We now turn to an analogous discussion of evidence in relation to the fourth characteristic, *reification of mathematical discourse*.

6.4 Reification of Mathematical Discourse

We also found a great variation in the ways in which participants engaged with the underpinning mathematical content in the task: in some responses such influence seemed *essential* in the diagnosing of student needs and in shaping intended pedagogical practice (e.g. where mathematical correctness or incorrectness *strengthens* or *restrains* intended practice); in some, such influence seemed *less essential*.

We note that in some scripts the respondents expanded substantially on underpinning mathematical content (e.g. on how Taylor's polynomial[3] is implicit in the discussion of tangency in the task) but explicitly reassured the reader that they do not advocate such advanced references in the classroom context of the task.

We characterised as *essential* (*restrained*) responses those where the respondent used a mathematically incorrect example (e.g. claiming that a certain line is a tangent of a certain function when it is not), or an inappropriate example (e.g. claiming that a certain example illustrates a certain feature of tangency when it does not). The responses from participants [3] and [8] are two such cases.

The response from participant [3] contains two examples of function graphs, $f(x) = \sqrt{|x|}$ and $f(x) = |x|$, which the participant attempts to present as illustrations, respectively, of one function that has a tangent at $x = 0$ and one that does not. She tries to explain this through resorting to mathematical theory. She includes in her writing the two graphs cited in Fig. 2:

We can explain to the student that, because $f'(x) = \dfrac{f(x)-f(x_0)}{x-x_0}$ and

ε: $y - f(x_0) = f'(x_0)(x - x_0)$
$y = f(x_0) + f'(x_0)(x - x_0) = g(x)$,

we have $\left|\dfrac{f(x)-f(x_0)}{x-x_0} - f'(x_0)\right| = \left|\dfrac{f(x)-f(x_0)-f'(x_0)(x-x_0)}{x-x_0}\right| = \left|\dfrac{f(x)-g(x)}{x-x_0}\right| = \tan(\omega)$

The line $g(x)$ is the tangent of the graph of $f(x)$ the more quickly the fraction $\dfrac{f(x)-f(x_0)}{x-x_0}$ goes to zero. In graph (2) $x \to x_0$ as quickly as $f(x) \to g(x)$ and so it is not a tangent [sic]

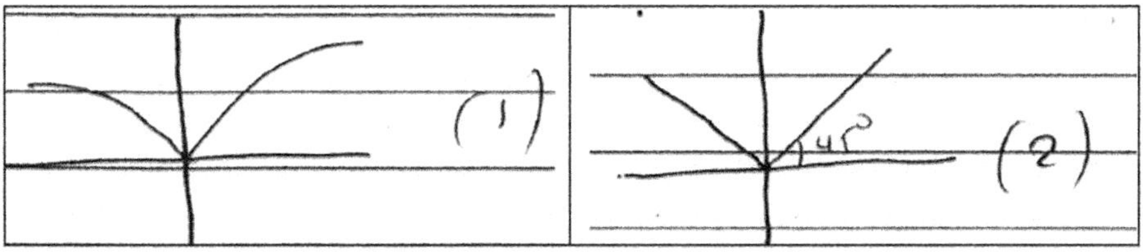

Fig. 2 Graphs (1) and (2) accompanying [3]'s claim "Why $f(x) = \sqrt{|x|}$ has a tangent at $x = 0$ but $f(x) = |x|$ doesn't"

[3] If f is a function such that there exists $f^{(k)}(x_0)$ *for every* $1 \leq k \leq n$, *the Taylor polynomial of degree* n *for* f *at* x_0 *is the polynomial* $T_{n,x_0}(x) = \sum_{k=0}^{n} \dfrac{f^{(k)}(x_0)}{k!}(x-x_0)^k$. It is the only polynomial (degree at most n) with the property $\lim_{x \to x_0} \dfrac{f(x)-T_{n,x_0}(x)}{(x-x_0)^n} = 0$ and it is the best polynomial approximation of f degree n at the point x^0. For $n = 1$ this polynomial approximation, $f(x_0)+f'(x_0)(x-x_0)$, is of degree one, therefore a line. The term *local straightness,* expresses this in visual terms: locally, the linear approximation of the curve and the curve itself are indistinguishable.

In the above excerpt, [3] claims that $f(x) = \sqrt{|x|}$ has a tangent at $x = 0$. But $f(x) = \sqrt{|x|}$ does not have a derivative at $x_0 = 0$ and so $f'(x_0)$ does not exist. Therefore the use of $f'(x_0)$ in the above formula is not possible. In sum, choosing this function as an example is irrelevant to the contrast participant [3] wishes to demonstrate. Furthermore, the explanation that aims to support the claim that $f(x) = |x|$ does not have a tangent at $x = 0$ is inscrutable (e.g. what does $f(x) \rightarrow g(x)$ mean?). Overall, we found the choice of examples inappropriate and the accompanying explanations confusingly presented, and in some parts of the script plainly incorrect.

The response from participant [8] attempts to describe how she would explain geometrically to the student that the tangent line is the limit of the secant lines. To this aim she proposes the images in Fig. 3, first for a circle, then for a general curve.

The text accompanying these images includes the following:

> …in this way we have managed that the students (a) realize the need for new knowledge (b) see visually the limit of the secant lines, namely the following:
>
> Since $\exists f'(x_o)\left[= \lim\limits_{x \to x_o} \dfrac{f(x)-f(x_0)}{x_0}\text{scribbled on top}\right]$ it means
>
> $\forall \varepsilon > 0, \exists \delta > 0 : 0 < |x - x_0| < \delta$ then
>
> $$\left|\frac{f(x)-f(x_0)}{x-x_0} - f'(x_0)\right| < \varepsilon \Leftrightarrow \left|\frac{f(x)-g(x)}{x-x_0}\right| < \varepsilon \quad (1)$$
>
> where $g(x) = f(x_0) + f'(x_0)(x - x_0)$, and so
> finally (1) $\Rightarrow |f(x) - g(x)| < \varepsilon|x - x_0|$ and in order to true for
> $x = x_0$
> $|f(x) - g(x)| \leq \varepsilon|x - x_0|$.

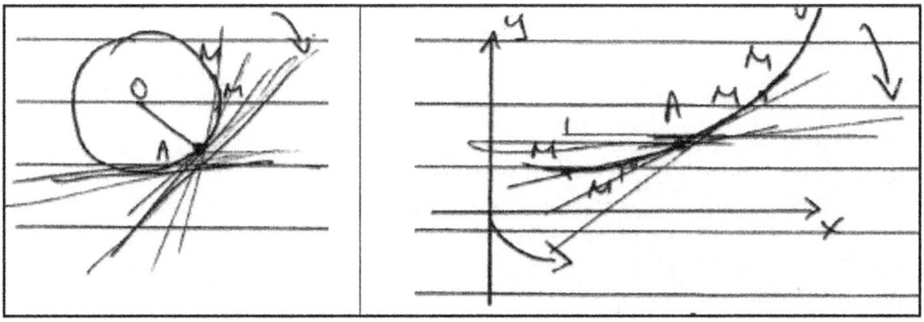

Fig. 3 The images proposed by [8] as an illustration of the tangent as the limit of secant lines (circle on the *left*, general curve on the *right*)

In the above excerpt, participant [8] uses the ε, δ definition to illustrate the formal relationship between the function and the tangent. However, she offers no explanation as to why and how this relation is evident here; or, how what she writes is connected with the previous images. In fact, the above formal presentation is a (dubious) attempt to describe local straightness, and not that the tangent is the limit of the secants. The latter can be described by the definition of the derivative, which presents the slope of the tangent line as the limit of the slope of the secant lines.

In contrast, the response from participant [12] is an example of a response in which engagement with the underpinning mathematical content in the task is *essential* and *strengthens* intended pedagogical practice: participant [12]'s response includes a reference to $T_1(x)$ (Taylor's polynomial for $n = 1$, the underpinning mathematical theory for local straightness):

$$\lim_{x\to x_0}\frac{f(x)-f(x_0)}{x-x_0}=f'(x_0)\Rightarrow \lim_{x\to x_0}\frac{f(x)-f(x_0)-f'(x_0)(x-x_0)}{x-x_0}=0$$

$$\Rightarrow \lim_{x\to x_0}\frac{f(x)-(f'(x_0)(x-x_0)+f(x_0))}{x-x_0}=0\Leftrightarrow \lim_{x\to x_o}\frac{f(x)-T_1(x)}{x-x_0}=0$$

This is accompanied by the presentation of a tangent line as "the line which approaches better the function's values". Subsequently, participant [12] attempts to express verbally what he sees as the relation between a function and its tangent at a point x_o (even though he does this in a global, rather than local sense). We see this as evidence in his response of substantially reified mathematical discourse – and, crucially, this reification is also robustly expressed in the resulting approach that he outlines. The script is written in a format resembling an imagined dialogue with the student in which participant [12] reconstructs the student's views of tangency through a carefully choreographed sequence of examples and counterexamples.

7 Discussion

By using the task in Fig. 1 we explored the participating 23 pre- and in-service teachers' competences in diagnosing issues pertaining to mathematics teaching situations and in addressing these issues in the feedback they intend to offer to students. We observed how participants weave together their views on mathematical, pedagogical and epistemological issues in order to diagnose key points in a teaching situation concerning the tangent line of a curve and then address these points in their intended practice. With illustrations from our analysis we proposed four characteristics – *consistency, specificity, reification of pedagogical discourse* and *reification of mathematical discourse* – as the theoretical lenses through which we can examine the participants' diagnosing and addressing of said teaching issues.

We sampled the insights that the analysis through these lenses can achieve in 6.1–6.4. In 6.1, for example, we continued the search initiated in the analyses in

Zachariades et al. (2013) to flesh out occasions in which even clearly stated beliefs are (or are not) attuned to the intended practice. In 6.2, we offered insight into the variable ways in which diagnosing and addressing issues pertaining to a particular teaching situation can be done with a sharp (or less so) focus on this very situation; and, we concluded that achieving a sharp and effective focus on the given situation can be a challenge. In 6.3 we exemplified how reification of key mathematics education theoretical constructs that participants had met in their postgraduate studies can exert a strong (or weak) influence on the articulation of coherent arguments. We indicated analogous influences with regard to the reification of mathematical discourse in 6.4.

In sum, we credit the identification of the insights in 6.1–6.4 to the potency of our four theoretical lenses – and of our situation-specific task design. We are aware, and heartened by the awareness, that the four characteristics we propose here accentuate issues that are raised and addressed in several other places in the related literature. For example, Mamolo and Pali (2014) deploy, inter alia, *Horizon Knowledge* (*HK*) (Ball et al., 2008) to investigate "the interplay between participants' personal solving strategies and approaches and their identified preferences when advising a student" (p. 32). We see our analysis with regard to *reification of mathematical discourse* as resonating with that of Mamolo and Pali.

Rowland and colleagues (Turner & Rowland, 2011) outline *Foundation* – one of the four features of the *Knowledge Quartet* (*KQ*) with the other three being *Connection*, *Transformation* and *Contingency* – as including, also inter alia, "overt subject knowledge, theoretical underpinning of pedagogy, use of terminology" (p. 200). Across our analyses through the four lenses, we are aiming to elaborate further the KQ's *Transformation* feature (which includes aspects such as "use of instructional material", "choice of representation" and "choice of examples", p. 200). Analogously to the links across our four lenses with *HK* (Ball et al., 2008), we see similar associations with the KQ's *Connection* feature (which includes "making connections between mathematical concepts", p. 201).

In line with Zazkis et al. (2013), the benefits of the task use we demonstrate in this chapter and the suggestion of the four characteristics "can be considered in three arenas: for teachers, for researchers, and for teacher educators" (p. 29). Engagement with tasks of these types "equips teachers with a repertoire of responses that they will be able to call upon in their "real teaching" (p. 29–30). To researchers and teacher educators, teachers' responses to these tasks "provide a window into imagined trajectories of 'good' teaching" (p. 30) – or otherwise – as well as a window onto "teachers' knowledge of mathematics, their knowledge of mathematics for teaching, as well as their pedagogical inclinations" (p. 30). Furthermore, through the proposed characteristics we can also assess the extent to which teachers have reified pedagogical and mathematical discourses that can shape the planning of their teaching, the implementation of this planning and their reflection on this implementation.

Overall, and in addition to the theoretical and methodological contribution we outline above, we propose our situation-specific task design and typology of four characteristics as potent components of formative and summative assessment in teacher education programmes. By accentuating the specificity of the classroom situation, we invite participants to reflect upon a peer's (the teacher in the task) approach as well as imagine their own intended practice. We thus gain insight into the participants' views (and, crucially, challenging aspects of these views) from a multi-faceted perspective (pedagogical, mathematical and epistemological, as initiated in our analyses of warrants that teacher ground their arguments, in Nardi et al., 2012). We see these four characteristics as a potent diagnostic tool for identifying the areas in which teachers' knowledge needs further support. As we observed, the analysis of the scripts in the terms of the four characteristics can offer substantial evidence of the learning outcomes achieved in the module of the post-graduate programme that these respondents are enrolled in. We note that, while in Sect. 4 we acknowledged certain inherent pitfalls of the insider-ness of our investigation, we however see this insider-ness as otherwise beneficial: for instance, we see in the scripts many of the discussions occurring first during the postgraduate programme being reproduced – sometimes well-assimilated, sometimes as fragmented, unsystematic regurgitations. These insights certainly impact on the way the programme is delivered in the future and the subsequent phases of our research also aim to encompass this impact. The subsequent phases of our research investigate this potency: for example, we are now supplementing the written responses to tasks with video-recorded group sessions in which participants reflect on the task, their responses to it and the responses of their peers.

Acknowledgement This study is partially supported by the grant of an annual *Erasmus Teaching Staff Mobility* programme that has been in place between our institutions since 2002.

References

Artigue, M., & Perrin-Glorian, M.-J. (1991). Didactic engineering, research and development tool: Some theoretical problems linked to this duality. *For the Learning of Mathematics, 11*(1), 3–17.

Ball, D., Thames, H. M., & Phelps, G. (2008). Content knowledge for teaching. *Journal of Teacher Education, 59*(5), 389–407.

Biza, I., Christou, C., & Zachariades, T. (2008). Student perspectives on the relationship between a curve and its tangent in the transition from Euclidean Geometry to Analysis. *Research in Mathematics Education, 10*(1), 53–70.

Biza, I., Nardi, E., & Joel, G. (2015). Balancing classroom management with mathematical learning: Using practice-based task design in mathematics teacher education. *Mathematics Teacher Education and Development, 17*(2), 182–198.

Biza, I., Nardi, E., & Zachariades, T. (2007). Using tasks to explore teacher knowledge in situation-specific contexts. *Journal of Mathematics Teacher Education, 10*, 301–309.

Biza, I., Nardi, E., & Zachariades, T. (2009). Teacher beliefs and the didactic contract on visualization. *For the Learning of Mathematics, 29*(3), 31–36.

Brousseau, G. (1997). *Theory of didactical situations in mathematics*. Dordrecht, The Netherlands/Boston, MA/London, UK: Kluwer Academic Publishers.

Carter, K. (1999). What is a case? What is not a case? In M. A. Lundeberg, B. B. Levin, & H. L. Harrington (Eds.), *Who learns what from cases and how? The research base for teaching and learning with cases*. Mahwah, NJ: Lawrence Erlbaum Associates.

Castela, C. (1995). Apprendre avec et contre ses connaissances antérieures: Un example concret, celui de la tangente. *Recherches en Didactiques des Mathématiques*, *15*(1), 7–47.

Charmaz, K. (2000). Grounded theory: Objectivist and constructivist methods. In K. Denzin & Y. Lincoln (Eds.), *Handbook of qualitative research* (pp. 509–535). Thousand Oaks, CA: Sage.

Christiansen, B., & Walter, G. (1986). Task and activity. In B. Christiansen, A.-G. Howson, & M. Otte (Eds.), *Perspectives on mathematics education: Papers submitted by members of the Bacomet Group* (pp. 243–307). Dordrecht, The Netherlands: D. Reide.

Dreher, A., Nowinska, E., & Kuntze, S. (2013). Awareness of dealing with multiple representations in the mathematics classroom – A study with teachers in Poland and Germany. In A. M. Lindmeier & A. Heinze (Eds.), *Proceedings of the 37th Conference of the International Group for the Psychology of Mathematics Education* (Vol. 2, pp. 249–256). Kiel, Germany: PME.

Erens, R., & Eichler, A. (2013). Belief systems' change – from preservice to trainee high school teachers on calculus. In A. M. Lindmeier & A. Heinze (Eds.), *Proceedings of the 37th Conference of the International Group for the Psychology of Mathematics Education* (Vol. 2, pp. 281–288). Kiel, Germany: PME.

Herbst, P., & Chazan, D. (2003). Exploring the practical rationality of mathematics teaching through conversations about videotaped episodes: The case of engaging students in proving. *For the Learning of Mathematics*, *23*(1), 2–14.

Hill, H. C., & Ball, D. L. (2004). Learning mathematics for teaching: Results from California's mathematics professional development institutes. *Journal for Research in Mathematics Education*, *35*(5), 330–351.

Leatham, K. R., Peterson, B. E., Stockero, S. L., & Van Zoest, L. R. (2015). Conceptualizing mathematically significant pedagogical opportunities to build on student thinking. *Journal for Research in Mathematics Education*, *46*(1), 88–124.

Leont'ev, A. (1975). *Dieyatelinocti, soznaine, i lichynosti [Activity, consciousness, and personality]*. Moskva, Russia: Politizdat.

Mamolo, A., & Pali, R. (2014). Factors influencing prospective teachers' recommendations to students: Horizons, hexagons, and heed. *Mathematical Thinking and Learning*, *16*(1), 32–50.

Markovits, Z., & Smith, M. S. (2008). Cases as tools in mathematics teacher education. In D. Tirosh & T. Wood (Eds.), *The international handbook of mathematics teacher education, Tools and processes in mathematics teacher education* (Vol. 2, pp. 39–65). Rotterdam, The Netherlands: Sense Publishers.

Mason, J., & Johnston-Wilder, S. (2006). *Designing and using mathematical tasks*. New York, UK: QED Press.

Nardi, E., Biza, I., & Zachariades, T. (2012). Warrant' revisited : Integrating mathematics teachers' pedagogical and epistemological considerations into Toulmin's model of argumentation. *Educational Studies in Mathematics*, *79*(2), 157–173.

Nardi, E., Healy, L., Biza, I., & Hassan AhmadAli Fernandes, S. (2016). Challenging ableist perspectives on the teaching of mathematics through situation-specific tasks. In C. Csíkos, A. Rausch, & J. Szitányi (Eds.), *Proceedings of the 40th Conference of the International Group for the Psychology of Mathematics Education (PME)* (Vol. 3, pp. 347–354). Szeged, Hungary: PME.

Sfard, A. (2008). *Thinking as communicating. Human development, the growth of discourse, and mathematizing*. New York, NY: Cambridge University Press.

Shulman, L. S. (1992). Toward a pedagogy of cases. In J. H. Shulman (Ed.), *Case methods in teacher education* (pp. 1–29). New York, NY: Teachers College Press.

Sierpinska, A. (1987). Humanities students and epistemological obstacles related to limits. *Educational Studies in Mathematics*, *18*, 371–397.

Sierpinska, A. (2003). Research in mathematics education: Through a keyhole. In E. Simmt & B. Davis (Eds.), *Proceedings of the Annual Meeting of Canadian Mathematics Education Study Group*. Acadia University. Wolfville, Nova Scotia: CMESG/GCEDM

Speer, M. N. (2005). Issues of methods and theory in the study of mathematics teachers' professed and attributed beliefs. *Educational Studies in Mathematics*, *58*(3), 361–391.

Tall, D., & Vinner, S. (1981). Concept image and concept definition in mathematics with particular reference to limits and continuity. *Educational Studies in Mathematics, 12*, 151–169.

Thompson, A. (1992). Teachers' beliefs and conceptions: A synthesis of the research. In D. A. Grouws (Ed.), *Handbook of research on mathematics teaching and learning* (pp. 122–127). New York, NY: Macmillan.

Tirosh, D., & Wood, T. (Eds.). (2009). *The international handbook of mathematics teacher education* (Vol. 2). Rotterdam, The Netherlands: Sense publishers.

Turner, F., & Rowland, T. (2011). The knowledge quartet as an organizing framework for developing and deepening teachers' mathematics knowledge. In T. Rowland & K. Ruthven (Eds.), *Mathematical knowledge in teaching* (pp. 195–212). London, UK/New York, NY: Springer.

Vinner, S. (1991). The role of definitions in the teaching and learning of mathematics. In D. Tall (Ed.), *Advanced mathematical thinking* (pp. 65–81). Dordrecht, The Netherlands: Kluwer Academic Publishers.

Vosniadou, S., & Verschaffel, L. (2004). Extending the conceptual change approach to mathematics learning and teaching. In L. Verschaffel & S. Vosniadou (Guest Eds.), *The conceptual change approach to mathematics learning and teaching, Special Issue of Learning and Instruction, 14*, 445–451.

Watson, A., & Mason, J. (2005). *Mathematics as a constructive activity: Learners generating examples*. Mahwah, NJ: Lawrence Erlbaum Associates.

Watson, A., & Mason, J. (2007). Taken-as-shared: A review of common assumptions about mathematical tasks in teacher education. *Journal of Mathematics Teacher Education, 10*(4–6), 205–215.

Zachariades, T., Nardi, E., & Biza, I. (2013). Using multi-stage tasks in mathematics education: Raising awareness, revealing intended practice. In A. M. Lindmeier & A. Heinze (Eds.), *Proceedings of the 37th Conference of the International Group for the Psychology of Mathematics Education (PME)* (Vol. 4, pp. 417–424). Kiel, Germany: PME.

Zaslavsky, O., & Sullivan, P. (Eds.). (2011). *Constructing knowledge for teaching: Secondary mathematics tasks to enhance prospective and practicing teacher learning*. New York, NY: Springer.

Zaslavsky, O., Watson, A., & Mason, J. (2007). Special issue: The nature and role of tasks in mathematics teachers' education. *Journal of Mathematics Teacher Education, 10*(4–6), 201–440.

Zazkis, R., Sinclair, N., & Liljedahl, P. (2013). *Lesson play in mathematics education: A tool for research and professional development*. New York, NY: Springer.

Diagnostic Competence for Dealing with Students' Errors: Fostering Diagnostic Competence in Error Situations

Hannah Heinrichs and Gabriele Kaiser

The project reported in this chapter aimed at assessing and fostering future teachers' diagnostic competence in teaching and learning situations in which students' errors occurred. Based on a model of diagnostic processes in error situations a university course and a pre- and post-test were developed. Probabilistic models (such as Item-Response-Theory and Latent-Class-Analysis) were used to assess the future teachers' diagnostic competence in erroneous teaching and learning situations based on their answers to the test items. The results show that the university course had an influence on the way future teachers hypothesize about the causes of the students' errors and partly also on their preferred way of dealing with errors. Additionally the importance of practical experience when fostering diagnostic competence in error situations became apparent.

1 Introduction

Errors are a necessary part of any learning effort. Therefore, mathematics teachers need to be able to diagnose and deal with errors made by students when learning mathematics. In order to support students' learning individually, teachers need diagnostic competence in several teaching situations. Errors occur in most learning processes and can enhance as well as inhibit understanding. Therefore, it is important for teachers to be able to deal with students' errors by being aware of their importance, knowing the reasons for errors and developing strategies to deal with errors.

In order to deal with errors in class it is helpful for teachers to learn about errors and diagnosis in their teacher education program. The study that is presented within

H. Heinrichs (✉) • G. Kaiser
University of Hamburg, Hamburg, Germany
e-mail: hannah.heinrichs@uni-hamburg.de

T. Leuders et al. (eds.), *Diagnostic Competence of Mathematics Teachers*,
Mathematics Teacher Education 11, DOI 10.1007/978-3-319-66327-2_4

this article set out to analyse to what extent the future teachers' diagnostic competence can be fostered within a university course in the first phase of German teacher education. To answer this question, the study was conducted at different universities in Northern Germany. Teacher education programmes in Germany differ greatly across universities. In this study, the relevant differences between the universities were taken into consideration and the results were interpreted accordingly.

This article presents the model that was used to conceptualize teachers' diagnostic competence in situations where students make errors (called "error situations" from here on). This model was then used to develop a university course to foster the university students' diagnostic competence in error situations. In a pre- and post-test design the future teachers' diagnostic competence in error situations was measured and then evaluated in terms of gains in this competence.

In the following, diagnostic competence is defined by taking a closer look at models of diagnostic processes and by introducing the model that was used in this study. On the basis of the theoretical considerations, the university course that was developed within the present study is presented and the methods used in the study are described. Furthermore, the results will be used to further analyze the nature of this conceptualization of diagnostic competence in error situations.

2 Diagnostic Competence in Error Situations

Teachers' diagnostic competence is an important prerequisite for successful teaching and has been focused on in several studies within the last years. Therefore, the discussion about diagnostic competence plays an important role in mathematics education. An important difference exists between approaches that focus on diagnostic competence as the "accuracy of judgement" (Schrader, 2006, p. 95) by comparing teachers' judgements with students' achievements in tests, and other approaches that focus more on diagnostic decisions that are made during teaching on an everyday basis (Abs, 2007; Praetorius, Lipowsky, & Karst, 2012). These decisions do not primarily strive to achieve accuracy but rather rely on the necessity to act in a teaching situation. Therefore, a more general notion defines diagnostic competence as all the abilities that are necessary to fulfill diagnostic tasks (Schrader, 2011, p. 683).

The present study focuses on diagnostic competence in error situations by taking a closer look at the competence that is necessary to analyze students' errors. This competence can be considered as part of the teachers' pedagogical content knowledge in the model of teachers' competence by Shulman (1986), who defines pedagogical content knowledge as "the ways of representing and formulating the subject that make it comprehensible to others" (Shulman, 1986, p. 9).

To model diagnostic competence, a general difference can be found in static and process models (Schrader, 1989). Static models are often used to assess diagnostic competence in the sense of the accuracy of judgement, whereas process models are used to divide the diagnostic process into several steps. In the present study a process model was developed to define and assess diagnostic competence in error situations and to develop an intervention to foster this competence.

Process models have been used in different contexts to model different processes that are associated with diagnostic decisions. These models can, for example, be found in teachers' decision-making, thought processes (Peterson & Clark, 1978), diagnostic processes (Jäger, 2010; Reisman, 1982), diagnostic teaching or formative assessment (Schoenfeld, 2011). These process models were taken into consideration when the model for the analysis of diagnostic competence in error situations was developed in this study.

In this study, diagnostic competence in error situations is defined as the competence that is necessary to come to implicit judgements based on formative assessment in teaching situations by using informal or semi-formal methods. The goal of this process is to adapt behaviour in the teaching situation by reacting to the student's error in order to help the student to overcome his/her misconception.

Diagnosing errors is an important part of the teaching process as can be seen in several studies on errors, for example in the study by Brown and Burton (1978) who claim that "one of the greatest talents of teachers is their ability to synthesize an accurate 'picture', or model, of a student's misconception from the meager evidence inherent in his errors. A detailed model of a student's knowledge, including his misconceptions, is a prerequisite to successful remediation" (Brown & Burton, 1978, S.155 f.). In order to use errors in the learning process, these errors have to be diagnosed individually to discover misconceptions and deal with them (Brown & Burton, 1978; Hußmann, Prediger, & Leuders, 2007; Putnam, 1987; Radatz, 1980; Schumacher, 2008).

However, before developing a model to conceptualize diagnostic competence in error situations, the term error has to be defined. In the German discussion about errors in learning processes in mathematics, errors are often regarded as statements that contradict general statements and definitions of mathematics or generally accepted methods (Heinze, 2004). In English several terms are used such as error, failure, slip and mistake (Seifried & Wuttke, 2010). One very common concept is the concept of misconceptions (Bell, 1993; Swan, 2001; Smith, diSessa, & Roschelle, 1993). These misconceptions are also called "alternative mathematical frameworks" and should not be considered as being wrong but rather as a transitional phase within the process towards fully understanding a concept. During this process learners often undergo different phases, which include generalizations that are not fully correct (Swan, 2001). In the following text the term "error" will be used, since it is the one that is most widely used when dealing with learning processes and teachers' diagnostic competence.

Since students' errors occur in teaching situations it is important to take a closer look at these situations. They vary significantly from one another, but they have in common that they are highly complex for the teacher who is diagnosing in these situations. The complexity of teaching situations was analyzed by Doyle (2006) and different characteristics of these situations have been highlighted to understand the complexity. On the one hand, teaching situations are multidimensional because different events and tasks take place at the same time and diverse people with different interests and goals are involved. The teacher has to help individual students and at the same time watch the whole class. In addition to that, teaching situations require immediate actions by the teacher and these actions have to take place quickly in order to keep up the momentum of the situation. Furthermore the situations are not predictable and any action by the teacher is public. All these aspects increase the complexity of teaching situations and actions taken by teachers within these situations.

To conceptualize the processes that teachers undergo to act and diagnose in these complex error situations, a model for diagnostic decisions in error situations was developed in the present study. This model is based on several theoretical considerations of teachers' decision-making and dealing with errors that will be presented in the following.

3 Modeling Teachers' Diagnostic Competence in Error Situations

Different approaches have been put forward in literature to model diagnosis in teaching situations. Many approaches focus on process models to analyze processes in teaching situations.

3.1 Modeling Diagnostic Processes

One important branch of research focuses on teachers' thought processes and analyses how teachers react in teaching situations when they perceive a discrepancy between their expectations and what is actually happening in class (Clark & Peterson, 1986; Peterson & Clark, 1978; Shavelson & Stern, 1981). These studies focus mainly on three areas of teachers' thought processes: in planning and during teaching as well as teachers' theories and beliefs. Especially teachers' interactive decisions (decisions during the teaching process) were analyzed within these studies. In models about teachers' interactive decisions the process always starts when the discrepancy between the teachers' expectations and the behaviour in class reaches a certain threshold. If teachers perceive this discrepancy to be intolerable, they usually look for alternative actions (Peterson & Clark, 1978). This model explains when teachers make decisions in teaching processes, however it does not explain how they react and whether these reactions are based on a diagnosis. To explain the processes between perceiving the discrepancy and reacting to it, different models have been put forward.

Rheinberg (1978) developed a model that consists of six phases. It starts with the "objective" students' behaviour, which has to be perceived by the teacher. The teacher then makes assumptions about the reasons underlying this behaviour. These assumptions influence the teacher's behaviour, which is then perceived by the learner and changes the learner's behaviour.

Another model that takes a more general look at diagnostic processes was developed by Klug, Bruder, Kelava, Spiel and Schmitz (2013). This model consists of three phases: the pre-actional, the actional and the post-actional phase. In the pre-actional phase, the teacher sets a goal, followed by actions to achieve that goal in the actional phase. These actions are systematically taken by gathering information to arrive at a diagnosis. In the post-actional phase, the pedagogical decisions derived from the diagnosis are implemented and communicated. As the diagnosis influences the behaviour in the next diagnostic situation, the pre-actional and the post-actional phases are closely linked.

These models were developed to refer to diagnostic situations in general. There are also models that take a closer look at diagnostic decisions in error situations in mathematics teaching.

Reisman (1982) developed the "diagnostic teaching cycle", which consists of five phases:

- "Identifying the child's weaknesses and strengths in arithmetic;
- Hypothesizing possible reasons for these weaknesses and strengths;
- Formulating behavioral objectives to serve as a structure for the remediation of weaknesses or the enrichment of strengths [...].;
- Creating and trying corrective remedial procedures [...];
- Continuing evaluation of all phases of the diagnostic cycle to see if progress is being made in either getting rid of trouble areas or in enriching strong areas" (Reisman, 1982, p. 5).

Reisman (1982) considers this model to be a cycle, since the remedial procedures have to be evaluated, thus starting the process from the beginning.

Cooper (2009) used a similar model consisting of three steps to develop a course in order to foster students' analysis of children's work to make instructional decisions. The three steps consist of identifying the error, making hypotheses about possible causes of the error and afterwards thinking about instructional strategies. This model is based on a model developed by Cooney (1988), which regards teaching as an interactive process consisting of "gathering information, making a diagnosis, and constructing a response based on that diagnosis" (Cooney, 1988, p. 273).

A similar approach was taken by Cox (1975), who states that after identifying an error, two questions are relevant: "(1) How can systematic errors be detected? And (2) once error patterns are identified, what methods can be used to remediate them?" (Cox, 1975, p. 151).

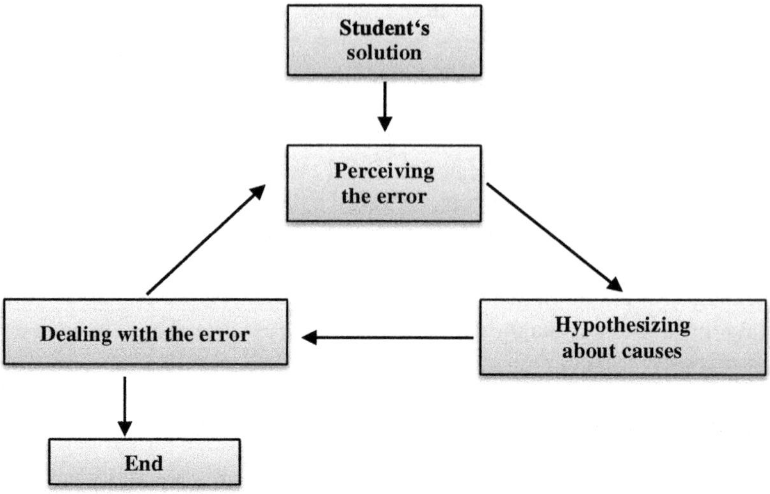

Fig. 1 Model of the diagnostic process in error situations (adapted from: Heinrichs, H. (2015). Diagnostische Kompetenz von Mathematik-Lehramtsstudierenden. Wiesbaden: Springer Spektrum, p.66. Copyright Springer Fachmedien Wiesbaden, with permission of Springer Nature)

3.2 Modeling Diagnostic Competence in Error Situations in the Present Study

In order to arrive at a model of the future teachers' diagnostic competence in error situations in the present study, the models presented above were taken into consideration by identifying steps which were relevant in each of the process models and which are considered to be relevant in error situations. Three steps could be found in all the models described above: perceiving or noticing, looking for reasons and then acting as a consequence.

Therefore, the model used in this study divides the process of teachers' diagnosis into three steps: First the error needs to be perceived and identified. Next, the teacher needs to hypothesize about the reasons for the error and on the basis of this hypothesis he/she needs to find an approach to deal with the error for the student to overcome the misconception (Fig. 1).

The first phase of perceiving the error is necessary in order to deal with the error afterwards. In some studies, this phase is already part of the definition of an error in learning processes as a reaction is only possible when an error is perceived (Seifried & Wuttke, 2010).

The second phase of the diagnostic process in error situations can be considered as the central part of the model as it can be found in all the models of diagnostic processes mentioned above. The importance of this phase is stressed by many researchers by referring to mostly theoretical considerations about learning processes and the importance of causes of errors within these processes.

For example, Ball, Thames and Phelps (2008) stressed the importance of finding the reasons for errors: "However, teaching involves more than identifying an incorrect answer. Skillful teaching requires being able to size up the source of a mathematical error. Moreover, this is work that teachers must do rapidly, often on the fly, because in a classroom, students cannot wait as a teacher puzzles over the mathematics himself" (Ball et al., 2008, p. 397).

It is especially important to identify the type of error, as stressed by Borasi (1996): "The nature of an error activity will also depend somewhat on the *type* of error considered, because different kinds of errors – such as incorrect definitions, correct results reached through incorrect procedures, wrong results, conjectures refuted by a counterexample, or contradictions, just to name a few significant categories – are likely to invite different kinds of questions for exploration and reflection" (Borasi, 1996, p. 280). In order to be able to identify the sources of errors, teachers need "to have a good hypothesis about what might be causing the error" (Ball, Hill, & Bass, 2005, p. 18). This is also important when correcting students' work, as Ashlock points out that "rather than just scoring papers, we need to examine each student's paper diagnostically – looking for patterns, hypothesizing possible causes and verifying our ideas" (Ashlock, 2010, p. 15).

All these statements already point at the reasons for errors being very diverse. There are always several different sources that can result in the same error. Therefore, different categorizations have been developed to differentiate between errors of different kinds. In this study errors were divided into categories, which were developed on the basis of typologies by Radatz (1980), Reisman (1982), Schoy-Lutz (2005), Cox (1975), Swan (2001), Tsamir (2005) and others. In the present study the future teachers' analysis of the causes of errors were divided into two categories: causes for an error to occur without considering the specific kind of error and causes for the specific error.

The causes of an error to occur can either be very general such as attention deficits or they refer to the task in a general way, for example problems in understanding the task or a general lack of knowledge in the topic.

The causes for the specific error cannot be described in a general way but refer to the specific error and can only be explained by looking at the error individually.

In the present study the competence to hypothesize about the causes of students' errors is defined as the ability to find different hypotheses about causes for one specific error and especially being able to name causes for the specific error and not only the general reasons for an error to occur. Additionally, people with a high level of this competence are able to identify plausible and implausible causes of an error.

The third phase of the diagnostic process in error situations is "dealing with the error" and can also be found in several models of diagnostic processes. There is a general consensus about the necessity to deal with errors in order to foster the learner's understanding but unfortunately there are only a few empirical studies on appropriate or effective ways of dealing with errors.

On the one hand the different approaches for dealing with errors can be distinguished by their tendency towards instructivist or constructivist theories. This can be found in a classification by Türling, Seifried and Wuttke (2012), who developed short video clips with different reactions by teachers in error situations, which amongst other aspects varied in "the extent to which the participants would give students hints for the correct solution" (Türling et al., 2012, p. 100). Son and Sinclair (2010) also used a similar classification by distinguishing between two approaches. On the one hand they identified approachs where the terms "show" and "tell" were used. These approaches were focused on visual and auditive presentations and therefore using instructivist methods. On the other hand they looked at approaches that were described by using the terms "give" and "ask", which included verbal or nonverbal requests for the student to get involved and thus stressing constructivist approaches towards learning.

Another differentiation that is based on the same idea is the one by Cooper (2009), who found a difference between teacher-directed instruction, where the teachers showed something to the students, and student–teacher interaction, where the teachers asked questions. This differentiation of approaches towards dealing with students' errors is closely linked to teachers' beliefs about learning and teaching mathematics, which are usually classified into transmissive and constructivist beliefs. In the classification of beliefs as well as in the classification of approaches towards dealing with errors, two ends of a continuum are considered and the theoretical assumptions about learning processes are similar for instructivist approaches to error situations and transmissive beliefs as well as for constructivist approaches in error situations and constructivist beliefs. Therefore, this classification of different approaches to error situation is also closely linked to the concept of "orientations" according to Schoenfeld (2011).

Besides the differentiation between instructivist and constructivist approaches, the ways of dealing with an error in a teaching situation can also be distinguished by the number of people involved in the process. Schoy-Lutz (2005) pointed out that teachers choose different reactions to deal with errors by either focusing on individual students or the whole group of students. Türling et al. (2012) also considered this differentiation by looking at whether teachers "take the entire class or single students into consideration by dealing with the shown problem/error" (Türling et al., 2012, p. 100). Therefore, these two aspects were considered in the phase of "dealing with the error" in the process model developed in this study.

This model of diagnostic competence was used to develop a university course to foster diagnostic competence in error situations and to assess this competence. In the following chapter, the university course that was developed within this study will be presented.

4 Fostering Diagnostic Competence in Error Situations

The present study's aim was to assess and foster future teachers' diagnostic competence in error situation during the first phase of teacher education. The intention of the study was to find out whether it is possible to foster the future teachers' diagnostic competence as well as identifying aspects that might enhance the development of diagnostic competences in future teachers.

Therefore, the study consisted of an intervention to foster teachers' diagnostic competence in error situations within a university course and a pre- and post-test to assess their diagnostic competence in error situations.

The intervention was conducted in four different universities in northern Germany with 138 future teachers participating in both pre- and post-test. The intervention consisted of four lessons of 90 min, which were conducted on a weekly basis.

The first lesson focused on hypothesizing about causes of students' errors. The second lesson dealt with possible reactions by the teacher. In the third lesson, the whole process was taken into consideration by applying it to different errors in arithmetic and in the fourth lesson errors in algebra were analyzed more closely as these errors are usually more complex.

Videos were used in the university course to present the error situations to the students. These videos were vignettes of 2 min in which students worked on tasks and made errors. To discuss ways of dealing with errors within the university course, in one video a teacher reacted to the students' error.

5 Development of the Pre- and Post-test and Analysis of the Data

The university course as well as the tests to assess the future teachers' diagnostic competence was developed on the basis of the model of diagnostic processes in error situations described above. The participants were introduced to the process model to use in error situations so that they could use this model in their analysis of errors in the post-test. However, they did not learn about specific errors that were addressed in the test. In the following, the design of the pre- and post-test is described. Afterwards the analysis of the data is specified and the hypotheses are stated.

Hypotheses

Three main hypotheses were proposed in the analysis of the data:

1. The future teachers' diagnostic competence in error situations is linked to other characteristics that were gathered in the questionnaire (such as beliefs about teaching and learning mathematics, practical experience, course of study, bachelor or master students, gender, university, and level of mathematics at school).
2. The university course has an effect on the future teachers' diagnostic competence in error situations.
3. The effect of the university course on the future teachers' diagnostic competence in error situations is influenced by other characteristics gathered in the questionnaire.

These hypotheses were tested in the analysis and the results will be presented afterwards.

5.1 Design of the Pre- and Post-test

The pre- as well as the post-test were based on four tasks, each of which dealt with one student's erroneous solution to a mathematical problem. Each of these tasks contained several items to assess the future teachers' diagnostic competence in error situations according to the process model.

The future teachers were assigned two tasks in the pre-test and two tasks in the post-test randomly, using a multi-matrix-design in order to make sure that they did not get the same task twice. Each task focused on one error. This error was analyzed by the future teachers in several items about each step in the diagnostic process model for error situations as described above. The design of the error tasks is illustrated by the following example, which uses a very common error in fractions.

Each task began with the error being presented and some information about the student's grade and mathematics class was given. The future teachers were then required to notice the error in the students' calculations by calculating another task and making the same error.

To analyze the future teachers' competence to hypothesize about causes of students' errors the future teachers were first asked to state possible reasons for the presented error in an open-ended response format. Afterwards they were given multiple-choice items and had to state whether the given causes are possible or impossible causes for that error (Fig. 2). These two approaches were chosen because it was assumed that not all future teachers will be able to come up with possible reasons by themselves but can rather recognize plausible causes when they are given in multiple-choice items. This also became apparent in the data analysis, which showed that the open items were more difficult than the multiple-choice items. Therefore the two methods of assessing the competence to hypothesize about

$\frac{1}{3} + \frac{2}{4} = \frac{3}{7}$	**Possible cause**	**Impossib-le cause**	**I don't know**
Sam did not read the task properly.			
Sam confused adding fractions with adding ratios.			
Sam considers the fraction line as a separation of two natural numbers.			
...			

Fig. 2 Multiple-choice item on the competence to hypothesize about causes of the error (Heinrichs, H. (2015). Diagnostische Kompetenz von Mathematik-Lehramtsstudierenden. Wiesbaden: Springer Spektrum, p.135. Copyright Springer Fachmedien Wiesbaden, with permission of Springer Nature)

	I would do this	I would probably do this	I probably wouldn't do this	I would not do this
I would tell him to check his calculation with the calculator.				
I would interrupt the students and explain the error to everyone.				
I would ask him to visualize his calculation.				
I would explain to him how to add fractions.				
...				

Fig. 3 Multiple-choice items on the preference when dealing with errors

causes of students' errors allowed to provide items of various levels of difficulty. This procedure was possible because the test was an online test and did not allow returning to the previous page.

After hypothesizing about the causes, the future teachers were asked to state their preferred way of dealing with the given error. These items were used to analyze the future teachers' preference when dealing with errors. Here, too, the future teachers first had to state their preferred reaction in open-ended items. On the next page they were given different ways of dealing with the error and had to state whether they would react that way or not on a four-point scale (Fig. 3).

In order to classify different approaches towards dealing with errors in the present study the suggested approaches to dealing with the error were systematically distinguished by their tendency towards instructivist or constructivist theories of learning and by whether the whole class or single students were taken into consideration. This way, a preference for a certain approach could be derived.

5.2 *Analysis of the Future Teachers' Answers*

The future teachers' answers to the open items in which they hypothesized about possible causes of the errors were first coded using qualitative content analysis according to Mayring (2010). For each item two people rated the answers using code manuals and arriving at high inter-coder reliabilities. These codes were assigned numbers from 1 to 3 depending on how specifically they referred to the error that was analyzed. To find out about the future teachers' competence to hypothesize about causes of students' errors, these codes and their answers to the multiple choice items were used to calculate latent variables by using Item Response Theory (IRT) analysis and showed an adequate fit (EPA/PV = 0.71).

The future teachers' answers to the open items about possible ways of dealing with the error were also coded by using qualitative content analysis. Then the future teachers' answers to the four-point-scale items were analyzed using latent class analysis. This method was chosen because the description of preferred ways of dealing with errors was a nominal description – there are not enough empirical findings to classify the different ways of dealing with errors according to their quality. The future teachers then were grouped into different classes with a similar preference in dealing with errors.

This way the future teachers' competence to hypothesize about causes of errors as well as their individual preference when dealing with errors could be reconstructed. In this analysis the future teachers' individual preferences when dealing with errors could be differentiated into three classes of teachers, who either preferred a constructivist or an instructivist approach or used both approaches flexibly. The constructivist approaches are characterized by more actively involving the student into the process of dealing with the error. These preferences were shown to be rather stable across different tasks dealing with different students' errors. The triangulation of the latent classes and the codes of the open items in the questionnaire showed that those future teachers who were classified as belonging to the class of students with a constructivist preference also asked for more students' involvement in the open items. This triangulation therefore supports the interpretation of the latent classes.

6 Results

When analyzing the future teachers' diagnostic competence and its relations to other characteristics before the university course to test the first hypothesis, it became apparent that practical experience (e.g. in the form of tutoring) is linked to the future teachers' competence to hypothesize about causes of students' errors (group comparison of students with/without practical experience: $t(134) = -2.077$, $p = 0.044$). Furthermore, the data suggests that mathematical content knowledge is associated with a higher competence level to hypothesize about causes of students' errors. However, mathematical content knowledge was not tested as such in specific items in the questionnaire, but future teachers who took a higher level mathematics course in their high school years showed a higher competence level when hypothesizing students' errors ($t(132) = 1.540$, p(1-tailed) $= 0.063$). As these future teachers had chosen this higher level course in high school voluntarily and were exposed to more complex mathematics than the ones who did not, it can be assumed that these students had a higher level of mathematical content knowledge. Additionally, the analysis showed that constructivist beliefs about learning and teaching mathematics are linked to a higher competence level when hypothesizing causes of students' errors ($r = 0.130$, $p = 0.067$). These links cannot be interpreted as causal influences but they can be seen as hints showing which aspects foster future teachers' diagnostic competence in error situations. They can also be used to test the conceptualization of diagnostic competence in error situations in this study as some of the links that were to be expected on a theoretical level were also found in the data.

When looking at the preference when dealing with errors in class, links could be found with the constructivist beliefs of the future teachers as higher constructivist beliefs about teaching and learning mathematics correlated with a higher preference for constructivist approaches to dealing with errors ($F(3132) = 4.912$, $p = 0.009$), which also supports the abovementioned interpretation of the latent classes. In addition to that master's students showed a higher preference for constructivist approaches.

Regarding the second hypothesis, the analysis of the future teachers' answers before and after the university course showed that the future teachers' diagnostic competence was influenced in both components (the competence to hypothesize about causes of errors and the individual preferences when dealing with errors) during the university course. The competence to hypothesize about causes of errors was significantly higher in the post-test than in the pre-test ($t(135) = -1.629$, p(1-tailed) = 0.05, Cohen's $d = 0.15$). The individual preference when dealing with errors in the post-test showed that slightly more future teachers preferred approaches that were more oriented towards constructivist teaching approaches (42 out of 137 future teachers preferred more constructivist approaches in the post-test). However, a lot of the future teachers showed the same preference in the pre- and in the post-test (76 showed the same preference when dealing with errors in the post-test as they did in the pre-test).

This means that the second hypothesis can be considered as true: On the one hand the future teachers showed a higher competence level when hypothesizing causes of students' errors. On the other hand, the future teachers showed a preference for more constructivist approaches to dealing with errors after having taken part in the university course than they did before. This is not necessarily better than the instructivist approach but this shows that this component of teachers' diagnostic competence in error situations can be influenced by a short university course.

In addition to the comparison of the diagnostic competence in the pre- and post-test the influence of other characteristics on the improvement of the future teachers' diagnostic competence was analyzed to test the third hypothesis. In this analysis no significant correlations with other characteristics could be identified, which can be considered as a hint of the university course being beneficial for different students. However, the results suggested that there is a group of future teachers who benefited strongly from the university course in both components of diagnostic competence in error situations. This group of future teachers might be very interested and motivated to deal with the topic.

7 Conclusion

To summarize, the present study succeeded in showing that the future teachers' diagnostic competence in error situations can be influenced positively in a short university course.

On the level of the future teachers' diagnostic competence in error situations the study showed that this competence is closely linked to the future teachers' content knowledge, practical experience, their beliefs and the progress in their studies. These links support the conceptualization of teachers' diagnostic competence in error situations in this study as these hypotheses about the links were developed on the basis of theoretical assumptions.

Concerning the preference when dealing with students' errors, the study showed that a classification regarding the preference of constructivist or instructivist approaches can be found and that this classification is linked to the future teachers' beliefs about teaching and learning mathematics, which supports the interpretation of this classification.

Regarding the effectiveness of the university course, this study showed that despite the short duration and complexity of the university course fostering diagnostic competences can indeed take place in university seminars.

However, it needs to be clarified that the study is a field study, because the university courses were conducted in the regular courses at the four universities without homogenizing the sample, choosing a representative sample or a control group. Also, as several hypotheses were tested in an explorative design, the multiple comparisons problem is to be dealt with. Furthermore, the study focused on mathematical concepts from lower secondary level and can therefore not easily be transferred to primary school or higher secondary school.

All in all, the study showed that future teachers' diagnostic competence can be fostered during university teacher education and that practical experience is helpful. It seems promising to implement courses on diagnostic competence into the regular teacher education programmes as well as add courses in which students' errors are analyzed more closely. Furthermore, it could be interesting to look at the links to other aspects of teachers' competence as assessed in several other studies.

References

Abs, H. J. (2007). Überlegungen zur Modellierung diagnostischer Kompetenz bei Lehrerinnen und Lehrern. In M. Lüders (Ed.), *Forschung zur Lehrerbildung* (pp. 63–84). Münster, Germany: Waxmann.

Ashlock, R. B. (2010). *Error patterns in computation: Using error patterns to help each student learn* (10th ed.). Boston, MA: Allyn & Bacon.

Ball, D., Hill, H. C., & Bass, H. (2005). Knowing mathematics for teaching. *American Educator, 29*(1), 14–17, 20–22, 43–46.

Ball, D., Thames, M. H., & Phelps, G. (2008). Content knowledge for teaching: What makes it special? *Journal of Teacher Education, 59*(5), 389–407.

Bell, A. (1993). Principles for the design of teaching. *Educational Studies in Mathematics, 24*, 5–34.

Borasi, R. (1996). *Reconceiving mathematics instruction: A focus on errors*. Norwood, OH: Ablex Pub.

Brown, J. S., & Burton, R. R. (1978). Diagnostic models for procedural bugs in basic mathematical skills. *Cognitive Science, 2*(2), 155–192.

Clark, C. M., & Peterson, P. L. (1986). Teachers' thought processes. In M. C. Wittrock (Ed.), *Handbook of research on teaching* (pp. 255–296). New York, NY: Macmillan.

Cooney, T. J. (1988). Teachers' decision making. In D. Pimm (Ed.), *Mathematics, teachers and children* (pp. 273–286). London, UK: Hodder und Stoughton in association with the Open University.

Cooper, S. (2009). Preservice teachers' analysis of children's work to make instructional decisions. *School Science and Mathematics, 109*(6), 355–362.

Cox, L. (1975). Diagnosing and remediating systematic errors in addition and subtraction computations. *The Arithmetic Teacher, 22*(2), 151–157.

Doyle, W. (2006). Ecological approaches to classroom management. In C. S. Weinstein, C. M. Evertson, & J. E. Brophy (Eds.), *Handbook of classroom management* (pp. 97–125). Mahwah, NJ: Lawrence Erlbaum.

Heinze, A. (2004). Zum Umgang mit Fehlern im Unterrichtsgespräch der Sekundarstufe *I. Journal für Mathematik-Didaktik, 25*(3/4), 221–244.

Heinrichs, H. (2015). Diagnostische Kompetenz von Mathematik-Lehramtsstudierenden. Wiesbaden: Springer Spektrum.

Hußmann, S., Leuders, T., & Prediger, S. (2007). Schülerleistungen verstehen –Diagnose im Alltag. *PM: Praxis der Mathematik in der Schule, 49*(15), 1–9.

Jäger, R. S. (2010). Diagnostische Kompetenz und Urteilsbildung als Element von Lehrprofessionalität. In O. Zlatkin-Troitschanskaia, K. Beck, D. Sembill, R. Nickolaus & R. Mulder (Eds.), Lehrerprofessionalität (S. 105–116). Landau, Germany: Verl. Empirische Pädagogik.

Klug, J., Bruder, S., Kelava, A., Spiel, C., & Schmitz, B. (2013). Diagnostic competence of teachers: A process model that accounts for diagnosing learning behavior tested by means of a case scenario. *Teaching and Teacher Education, 30*, 38–46.

Mayring, P. (2010). *Qualitative Inhaltsanalyse: Grundlagen und Techniken* (11th ed.). Weinheim, Germany: Beltz, Beltz Pädagogik.

Peterson, P. L., & Clark, C. M. (1978). Teachers' reports of their cognitive processes during teaching. *American Educational Research Journal, 15*(4), 555–565.

Praetorius, A.-K., Lipowsky, F., & Karst, K. (2012). Diagnostische Kompetenz von Lehrkräften: Aktueller Forschungsstand, unterrichtspraktische Umsetzbarkeit und Bedeutung für den Unterricht. In R. Lazarides (Ed.), *Differenzierung im mathematisch-naturwissenschaftlichen Unterricht* (pp. 115–146). Bad Heilbrunn, Germany: Klinkhardt.

Putnam, R. T. (1987). Structuring and adjusting content for students: A study of live and simulated tutoring of addition. *American Educational Research Journal, 24*(1), 13–48.

Radatz, H. (1980). *Fehleranalysen im Mathematikunterricht.* Braunschweig, Germany: F. Vieweg.

Reisman, F. K. (1982). *A guide to the diagnostic of teaching of arithmetic* (3rd ed.). Columbus, OH: C. E. Merrill.

Rheinberg, F. (1978). Gefahren Pädagogischer Diagnostik. In K. J. Klauer (Ed.), *Handbuch der pädagogischen Diagnostik* (pp. 27–38). Düsseldorf, Germany: Schwann.

Schoenfeld, A. H. (2011). Toward professional development for teachers grounded in a theory of decision making. *ZDM Mathematics Education, 43*(4), 457–469.

Schoy-Lutz, M. (2005). *Fehlerkultur im Mathematikunterricht: Theoretische Grundlegung und evaluierte unterrichtspraktische Erprobung anhand der Unterrichtseinheit "Einführung in die Satzgruppe des Pythagoras"*. Hildesheim, Germany: Franzbecker.

Schrader, F.-W. (1989). *Diagnostische Kompetenzen von Lehrern und ihre Bedeutung für die Gestaltung und Effektivität des Unterrichts.* Frankfurt, Germany: Lang.

Schrader, F.-W. (2006). Diagnostische Kompetenz von Eltern und Lehrern. In D. H. Rost (Ed.), *Handwörterbuch Pädagogische Psychologie* (pp. 68–71). Weinheim, Germany: Beltz.

Schrader, F.-W. (2011). Lehrer als Diagnostiker. In E. Terhart (Hrsg.), *Handbuch der Forschung zum Lehrerberuf* (S. 683–698). Münster, Germany: Waxmann.

Schumacher, R. (2008). Der produktive Umgang mit Fehlern: Fehler als Lerngelegenheit und Orientierungshilfe. In R. Caspary (Ed.), *Nur wer Fehler macht, kommt weiter* (pp. 49–72). Freiburg im Breisgau, Germany: Herder.

Seifried, J., & Wuttke, E. (2010). "Professionelle Fehlerkompetenz" – Operationalisierung einer vernachlässigten Kompetenzfacette von (angehenden) Lehrkräften. *Wirtschaftspsychologie, 12*(4), 17–28.

Shavelson, R. J., & Stern, P. (1981). Research on teachers' pedagogical thoughts, judgements, decisions and behavior. *Review of Educational Research, 51*(4), 455–498.

Shulman, L. (1986). Those who understand: Knowledge growth in teaching. *Educational Researcher, 15*(2), 4–14.

Smith, J. P., diSessa, A. A., & Roschelle, J. (1993). Misconceptions reconceived: A constructivist analysis of knowledge in transition. *The Journal of the Learning Sciences, 3*(2), 115–163.

Son, J.-W., & Sinclair, N. (2010). How preservice teachers interpret and respond to student geometric errors. *School Science and Mathematics, 110*(1), 31–46.

Swan, M. (2001). Dealing with misconceptions in mathematics. In P. Gates (Ed.), *Issues in mathematics teaching*. London, UK: Routledge Falmer.

Tsamir, P. (2005). Enhancing prospective teachers' knowledge of learners' intuitive conceptions: The case of same A–same B. *Journal of Mathematics Teacher Education, 8*(6), 469–497.

Türling, J. M., Seifried, J., & Wuttke, E. (2012). Teachers' knowledge about domain specific student errors. In J. Seifried & E. Wuttke (Eds.), *Learning from errors at school and at work* (Vol. 1, pp. 95–110). Opladen, Germany: Budrich.

Factors Influencing the Accuracy of Diagnostic Judgments

Andreas Ostermann

Diagnostic judgments suffer from several biases, one of which is the so-called 'expert blind spot'. Looking at this phenomenon from a cognitive perspective helps to explain certain tendencies to misjudge students. The chapter reports empirical results about factors influencing diagnostic judgments from mathematics education and discusses the hypothesis that teachers underestimate the difficulty of tasks resp. overestimate the performance of students.

1 Introduction

In order to teach adaptively, teachers must be able to take students' perspectives into account. However, research shows that teachers tend to overestimate students' performance (e.g., Nathan & Koedinger, 2000), which may be explained by the so-called *expert blind spot*: It is hypothesized that teachers are not able to assume students' perspectives due to their extensive subject matter knowledge (Nathan & Petrosino, 2003). Furthermore, when teachers estimate the difficulty of a task, knowledge of higher mathematics might cause an illusion of simplicity (c.f. Kelley, 1999). To achieve an appropriate judgment in accordance with the students' thinking, other factors like knowledge of students' typical cognitions may help to adjust the first guess.

This chapter discusses a variety of factors that influence the accuracy of teachers' judgments. The accuracy is traditionally used as a measure of teachers' diagnostic competence (Helmke & Schrader, 1987; Südkamp, Kaiser, & Möller, 2012; see also chapter "Diagnostic Competence of Mathematics Teachers – Unpacking a Complex Construct" in this book). Although there is a large strand of research on such influencing factors, there is no coherent model that classifies these factors

A. Ostermann (✉)
University of Education Freiburg, Freiburg, Germany
e-mail: andreas.ostermann@ph-freiburg.de

T. Leuders et al. (eds.), *Diagnostic Competence of Mathematics Teachers*,
Mathematics Teacher Education 11, DOI 10.1007/978-3-319-66327-2_5

sufficiently. In this chapter, the nature and the extent of influencing factors is exemplified based on empirical findings and theoretical considerations.

An example might help to show how multifarious the influences on diagnostic judgments can be: Fig. 1 illustrates a task in the area of functions and graphs which is typical for 8th grade high school curriculum in Germany (Leuders, 2017). The judgment of the difficulty of the task can be performed by estimating the percentage of students (from an average 8th grade high school class) that will presumably solve this task correctly.

A teacher may draw on many facets of professional knowledge in order to appropriately determine the difficulty of the task. Firstly, the estimated difficulty depends on the curricular status of the class. More specifically, it relies on mathematical concepts that are part of school-related subject matter knowledge, for example, the definition of a function, the aspects *assignment*, and *covariation* pertaining to the concept of function (Vollrath, 1989), the use of coordinate systems or charts. These aspects of the task related to the assumed knowledge of the students might be factors that determine the difficulty of the task and refer to conceptual aspects that can be defined purely by mathematical analysis. Furthermore, the task requires the interpretation of the given situation by means of mathematical or physical concepts; in this case, the concept of velocity. Here mistakes may occur that rather originate from (typical) students' thinking relying on everyday experience than from mathematical analysis. With respect to this category, research has found a diversity of students' misconceptions (Leinhardt, 1990; Hadjidemetriou & Williams, 2002). For instance, students who tend to sketch graphs generally as continuous, discontinuity

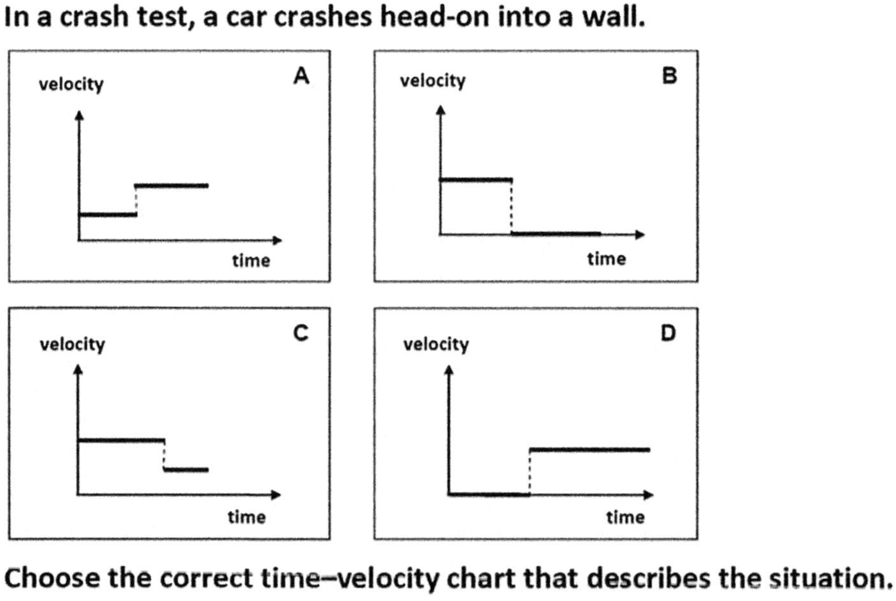

Fig. 1 Which percentage of an average 8th grade high school class will presumably solve this task correctly? (Leuders, 2017; Ostermann, Leuders, & Philipp, 2017)

might cause difficulties. Moreover, some students tend to misinterpret the graphs as a picture of the underlying situation. Thus, students might erroneously choose option D, which might suggest a wall at the discontinuous place. If teachers are unaware of this *graph-as-picture-error* or the *continuous prototype,* they will probably underestimate the task's difficulty. Knowledge of students' typical misconceptions, such as in the examples mentioned, cannot be derived from the underlying pure mathematical concepts.

To classify the facets of knowledge that may be used in diagnostic judgments, one can refer to the model of pedagogical content knowledge (PCK) of Ball, Thames and Phelps (2008). Although Ball et al. (2008) classified the factors based on primary school teachers' knowledge, the taxonomy seems to be useful to distinguish areas of teachers' knowledge of other school types as well (see Fig. 2). The knowledge of the mathematical complexity of a task can be seen as a part of specialized content knowledge (SCK) and knowledge of content and curriculum (KCC). The SCK is defined as school-related subject matter knowledge that is only needed by teachers and not by mathematicians in other professions. Highly developed SCK contains knowledge of different ways to solve a task and the relevant solution steps, and thus seems to be a necessary prerequisite to estimate the difficulty of a task. Knowledge of content and students (KCS), which contains knowledge of students' misconceptions and students' typical strategies or cognitions, forms another important diagnostic component of teachers' professional knowledge. Further, horizon content knowledge, which includes the knowledge of higher mathematics acquired at university, might cause misjudgments in the sense of the

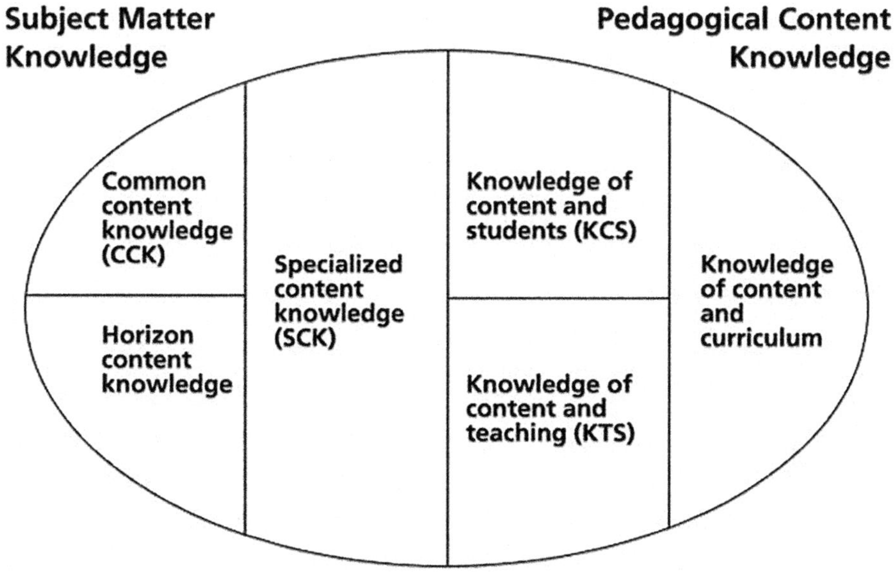

Fig. 2 Mathematical knowledge for teaching (Ball et al., 2008, p. 403)

expert blind spot (e.g., in the example of Fig. 1 teachers might not be aware of students' misconceptions and thus overestimate the expected solution rate).

When considering the influence on diagnostic competence later in this chapter, we will concentrate on the role of horizon content knowledge and knowledge of students' misconceptions (as part of *KCS*). Furthermore, we will see that diagnostic decisions are not only influenced by facets of teachers' professional *knowledge,* but also by outer factors such as *reference groups* or the *setting or context* in which judgments are made (cf. also Klug, Bruder, Kelava, Spiel, & Schmitz, 2013). We will discuss consequences of social biases and judgment settings on the operationalization of diagnostic competence. These considerations will lead us to reflections about dual-process models and heuristic decision making: A model that Nickerson (1999) devised to describe the knowledge of experts about laypersons' knowledge will be introduced at the end of this chapter to provide a framework for the process of diagnostic judgments and an explanation of teachers' tendency to overestimate students' performance.

2 Factors Influencing Diagnostic Judgments

2.1 Models of Factors Influencing Diagnostic Judgment Accuracy[1]

As we have seen in the introduction, teachers' ability to judge the difficulty of tasks appropriately depends on many factors, whose influence is worthwhile to investigate. A large quantity of studies were conducted within the so-called veridicality paradigm, in which the accuracy of judgments is considered as an indicator for underlying competences. According to Hoge and Colardaci (1989) different operationalizations of judgment accuracy show different degrees of specificity:

> (a) ratings (low specificity), where teachers rated each student's academic ability (e.g., "lowest fifth of class" to "highest fifth of class"); (b) rankings, where teachers were asked to rank order their students according to academic ability; (c) grade equivalence, where teachers estimated, in the grade-equivalent metric, each student's likely performance on a concurrently administered achievement test; (d) number correct, where teachers were asked to estimate, for each student, the number of correct responses on an achievement test, administered concurrently; and (e) item responses (high specificity), where teachers indicated, for each item on an achievement test administered concurrently to the students, whether they thought the student would respond correctly to the item or had sufficient instruction to respond correctly (Hoge & Colardaci, 1989, S.300f).

The meta-analyses of Südkamp et al. (2012) and Hoge and Colardaci (1989) compare data of the rank-order-accuracy and show mean correlations of $r = 0.66$ and $r = 0.63$, respectively. The results of Hoge and Colardaci point out that teachers perform

[1] Section 2.1 is taken from Ostermann, Leuders and Philipp (2017, in press) and translated by the author.

better in direct judgments (e.g., a judgment related to a specific presented task) than in indirect judgments (which rather are related to students' general abilities).

Significant results could be only found in the dimension of judgment characteristics: Just as Hoge and Colardaci (1989), Südkamp et al. found that teachers perform better in direct judgments than in indirect judgments (Fig. 3).

The rank correlations mentioned above may seem to indicate that the teachers' judgments have a satisfactory quality. However, one has to bear in mind that correlations cannot measure systematic over- or underestimations. If, for instance, a teacher systematically overestimates students' performance by a certain percentage, this will not influence the rank correlation. Thus, by investigating only the rank correlation, teachers' tendencies to over- or underestimate students' performance cannot be detected as a significant *teacher characteristic*. Since systematical overestimations (e.g., Nathan & Koedinger, 2000; Spinath, 2005) and underestimations (Selter, 1995) are well documented in mathematical education research, one should include the analysis of studies that also investigated teachers' estimation of solution rates.

In order to develop a theory about teachers' cognitive processes, one should directly investigate the influence of certain facets of knowledge (in the sense of Mathematical Knowlegde for Teaching (MTK)) on diagnostic competence. The fact that direct judgments lead to better accuracies could be a hint that teachers actually draw on specific knowledge of students' typical solution processes related to the specific task (c.f. Ostermann et al., 2017).

While task-specificity plays a key role in the two meta-studies above, there were no meta-analytic findings on the role of group specificity. In this respect, the factors discussed in the Section 2.2 complement the perspective of Südkamp et al. (2012). In the area of judgment characteristics, the role of social biases and their consequence on the operationalization of diagnostic judgments is discussed.

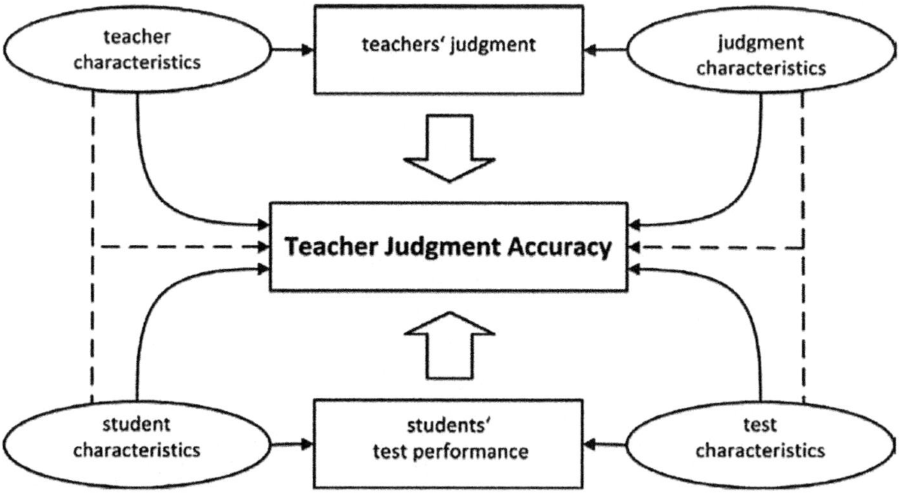

Fig. 3 A model of teacher-based judgments of students' academic achievement (Südkamp et al., 2012)

2.2 Social Biases and Group Specificity

Recent studies on the accuracy of teacher judgments predominantly refer to specific student groups (e.g., Baumert et al., 2010; Krauss et al., 2008). One disadvantage of this approach is that special characteristics of specific groups may bias judgments. Thus, the statistical comparison of different teachers' judgment accuracy within one study can be biased by confounding variables.

For example, Dünnebier, Gräsel and Krolak-Schwerdt (2009) report types of anchoring, according to which teachers' judgment accuracy is influenced by marks given by other teachers. The estimated number of correct solutions also depends on the reference group, particularly high-performing versus low-performing groups (Südkamp & Möller, 2009). Kaiser, Helm, Retelsdorf, Südkamp and Möller (2012) identified further anchors referring to individual students, such as information about intelligence or performance in other school subjects.

The estimation of specific groups may provide a high extent of ecological validity, but suffers – as already stated – from biases by a range of confounding variables. In order to statistically investigate factors that influence teachers' judgment accuracy, the estimation of many teachers as dependent variable must be comparable and thus, in order to avoid confounders, it would be necessary that all the investigated teachers rate the same student group. However, a sufficiently large sample of teachers, who are all familiar to the rated student group to an equal extent, seems hard to achieve practically.

These reflections suggest that it may be useful to distinguish between components of diagnostic competence which relate to a specific student group versus a non-specific student group. For example, Karst, Schoreit and Lipowsky (2014) investigate class-referred, a student-global and a student-specific diagnostic competence; whereas, class-referred diagnostic competence means a task-related competence and the other two components are related to persons (groups and individuals).

The aforementioned biases can cause problems within the veridicality paradigm. One can face these problems by abstaining specific groups and therefore investigating judgments related to "typical" or "representative" student groups (c.f. Karst, 2014; Ostermann et al., 2017). In order to measure the accuracy of teachers' judgments of non-specific groups, the components of judgment accuracy, traditionally investigated in recent research, must be adjusted as described in Table 1. This approach, however, requires data from a typical or representative sample of students. Solution rates of a representative sample of students provide the *empirical task difficulty*.

These components can still be seen as indicators for diagnostic competence as they integrate several factors that determine the difficulty of a task and their weighting; however, only for non-specific student groups (as illustrated in the example in Fig.1). In this approach the focus is on knowledge of task-specific requirements. This knowledge can be considered a prerequisite for student-specific diagnostic competence, which requires the fitting between task requirements and student-abilities (cf. Karst, 2012).

Table 1 Another approach to operationalize diagnostic competence: Measuring the accuracy of teachers' judgments of non-specific student groups

Rating of specific student groups	Rating of non-specific student groups
Solution-rate accuracy: Teachers' estimation of the number of correct responses within a specific student group referring to a specific task can be compared with the actual number of students' correct responses. The difference between the estimated and the actual number of correct responses is defined as solution-rate-accuracy	*Solution-rate accuracy:* Teachers' estimation of the number of correct responses within a non-specific student group referring to a specific task can be compared with the empirical solution rate of a representative or typical student sample. The difference between the estimated and the empirical percentage of correct responses is defined as solution-rate-accuracy
Rank-order accuracy (Students' Performance): Teachers' estimated rank-order of students according to their performance can be compared with the actual rank-order of students. The correlation between the estimated and the actual rank-order is defined as rank-order-accuracy	*Rank-order accuracy (Task Difficulties):* Teachers' estimated rank-order of several tasks according to their difficulty can be compared with the *empirical rank-order* of tasks which is determined by the performance of representative or typical student samples. The correlation between the estimated and the empirical rank-order defines the rank-order accuracy of task difficulties

Nevertheless, Spinath (2005) pointed out that the traditional components of the *veridicality paradigm* (Helmke & Schrader, 1987) are empirically independent. This raises the question which sub-skills or facets of knowledge were represented in each component. Therefore, in the next sections, we are looking at such facets of knowledge in the sense of MKT and their relation to diagnostic competence.

2.3 Horizon Content Knowledge

This paragraph illustrates how mathematical subject matter knowledge causes the overestimation of students' performance. Nathan & Koedinger (2000) showed that both, teachers with professional experience and pre-service teachers, tend to show misjudgments in the estimation of the difficulty of given tasks: Participants were asked to estimate the difficulty of algebraic and arithmetic tasks with regard to students. Results showed that teachers' estimations were not in line with the empirical rank-order of the tasks difficulty, according to which students perform better in everyday-language-tasks than in formal language tasks. High-school-teachers, who studied higher mathematics more intensively than primary and secondary school teachers, showed the largest misjudgments. The authors interpret their findings as a symptom of the so-called expert blind spot: Extensive subject matter knowledge impedes teachers from taking students' perspectives (Nathan & Petrosino, 2003). Generally speaking, the experts' tendency to overestimate laypersons knowledge seems to be a ubiquitous phenomenon that is not limited to mathematics teaching (Chi, Siler, & Jeong, 2004; Herppich, Wittwer, Nückles, & Renkl, 2010; Hinds, 1999;

Lin & Chiu, 2010). The term *Expert-Blind-Spot* refers to experts' difficulty to put themselves in the position of laypersons and anticipate their difficulties in acquiring professional contents.

A recent study showed that teachers' estimation of students' performance in tasks in the area of functions and graphs correlates with mathematicians' own effort to solve the task (Ostermann, Leuders & Nückles, 2015). Especially, the solution rates of graphical tasks were overestimated to a higher degree than numerical tasks. We interpret this result as a phenomenon of knowledge-encapsulation (c.f. Rikers, Schmidt, & Boshuizen, 2000). In solving graphical tasks, teachers may percept all relevant information at a glance; whereas, numerical tasks require several calculations steps, which teachers might experience as an effort to solve the task on their own. Possibly teachers are not able to anticipate different solution steps of graphical tasks or evaluate their difficulty correctly with regard to students (Ostermann et al., 2015).

2.4 Knowledge of Content and Students

Regarding diagnostic competence as the teachers' ability to accurately estimate students' characteristics (Schrader, 2006), *knowledge of content and students* must be seen as a diagnostic key facet of *mathematical knowledge for teaching*. The question if KCS in the area of functions and graphs influences the accuracy of pre-service teachers' diagnostic judgments of non-specific student groups was investigated in a recent intervention study (Ostermann et al., 2017). Results show that a short intervention, which provides knowledge of students' misconceptions, improves both components, the solution-rate accuracy and the rank-order accuracy; whereas, a pure sensitization for teachers' tendency of misjudgments only improves the solution-rate accuracy.

These findings are in line with the work of Spinath (2005) with regard to the independency of the components, but suggest further interpretation: Sensitization seems to be a rather superficial intervention, which only leads pre-service teachers to constantly reduce of the estimated percentage solution rate to a certain amount, trying to not systematically overestimate students' performance. In contrast, knowledge of students' misconceptions helps pre-service teachers to reflect substantially on the difficulty of given tasks. Ranking tasks according to their difficulty requires reflection of several difficulty-creating task characteristics and thus, the integration of both, mathematical concepts (SCK) and characteristics beyond pure mathematics, such as students' misconceptions (part of KCS).

2.5 Decompression of Tasks

While in the taxonomy of Ball et al. (2008) the facet of *KCS* is highly associated with diagnostic knowledge, Morris, Hiebert and Spitzer (2009) emphasize the role of school-related subject matter knowledge (SCK). In order to foster teachers' diagnostic competences, the authors propose a training of decompression of tasks into solution steps. Using tasks in the domain of fractional arithmetic, they investigated teacher students' ability to dissect (resp. decompress) tasks in their elementary sub-goals. An ideal decompression of the addition of two fractions is proposed by the authors as shown Philipp (2017, chapter "Diagnostic Competences of mathematics teachers with a view to processes and knowledge resources" in this book). According to Morris et al. (2009) such a decompression contains all relevant steps and concepts which must be acquired by students in order to solve a task completely. Thus, teachers' ability to decompress can be a helpful factor in diagnostic situations for identifying concepts that students did not understand sufficiently, and identifying misconceptions that must be corrected. Therefore, teacher students should be provided with learning opportunities which explicitly train the decompression of tasks.

It is questionable, however, to what extent such a linear understanding of decompression can be transferred to other areas, such as functions and graphs (e.g., Fig. 1). In this case, decompression might be defined as knowledge of all mathematical concepts required to understand and solve the task. However, then it seems to be artificial to detect students' misconceptions by the means of decompression.

2.6 Deliberative and Intuitive Decisions

If teachers estimate the difficulty after decompression or further analysis of a task, the judgment is not anymore an intuitive but a deliberative decision (Betsch, 2004; Gigerenzer & Gaissmaier, 2011). The distinction between intuitive and deliberative judgments has become more popular in judgment research by the heuristics-and-biases program and led to the well-known Dual-Process-Theories (Evans, 2008; Kahneman, 2003; Tversky & Kahnemann, 1974).

Numerous studies show that different decision modes lead to different judgment quality and satisfaction (e.g., Gigerenzer & Gaissmaier, 2011; Plessner & Czenna, 2008; Wilson & Schooler, 1991). Due to their implicit knowledge, experts in a domain seem to judge more appropriately in an intuitive decision mode, because a deliberative-decision mode might lead to an overvaluation of irrelevant aspects (Plessner & Czenna, 2008). In a study of Plessner, Freiberger, Kurle and Ochs (2006) soccer experts' predicted the outcome of the Fédération Internationale de Football Association (FIFA) confederation cup 2005 better in the early-and-intuitive mode than in late-and-analytic mode, as later a lot of information becomes available that could be reflected on. According to Gigerenzer and Gaissmaier (2011), the reduction to little but relevant information can improve judgment quality: "Ignoring part of the information can lead to more accurate judgments than weighting and adding all information" (ibid., S.451).

According to the author's best knowledge, there are no studies on mathematics teachers' diagnostic decisions that explicitly investigate the influence of intuitive versus deliberative modes of decision on judgment accuracy. It is quite conceivable that certain operationalizations in studies (e.g., think aloud interviews) induce a deliberative decision modus, which might influence the judgment accuracy. However, taking findings of recent expertise research into account, it would be logical to assume that prospective teachers perform better in a deliberative than in an intuitive mode, because as diagnostic novices they can rely only on little implicit knowledge. On the other hand, experienced teachers should perform better in an intuitive or heuristic modus of decision due to their implicit knowledge. Regarding the requirements of everyday classroom practice, teachers should be able to decide quickly and intuitively.

Considering the use of heuristics, we present a heuristic model on the process of estimation of other persons' knowledge in the next section, which seems to be applicable to describe diagnostic situations.

3 A Heuristic for the Process of Diagnostic Judgments

Within the research paradigm of dual process models, theories of "decisions under uncertainty" were developed. One very famous heuristic is the so-called *anchoring and adjustment heuristic* (Tversky & Kahneman, 1974), which states that people intuitively take well-available anchors as initial points for decisions or estimations, and refine those estimations by situational appropriate adjustments. These anchors can sometimes be the reason for misjudgments or erroneous projections (Kahneman, 2003; Nickerson, 1999); especially, when they dominate the decision and were not adjusted appropriately.

Nickerson (1999, 2001) proposes a model based on an anchoring and adjustment heuristic (see also Leuders, Dörfler, Leuders, & Philipp, 2018, chapter "Diagnostic Competence of Mathematics Teachers – Unpacking a Complex Construct" and Philipp 2017, chapter "Diagnostic Competences of mathematics teachers with a view to processes and knowledge resources" in this book). The model describes how people (intuitively) assume what specific other people know (illustrated in Fig. 4). According to Nickerson, people who estimate other peoples' knowledge always start with their own knowledge as initial anchor. This anchor is subsequently refined

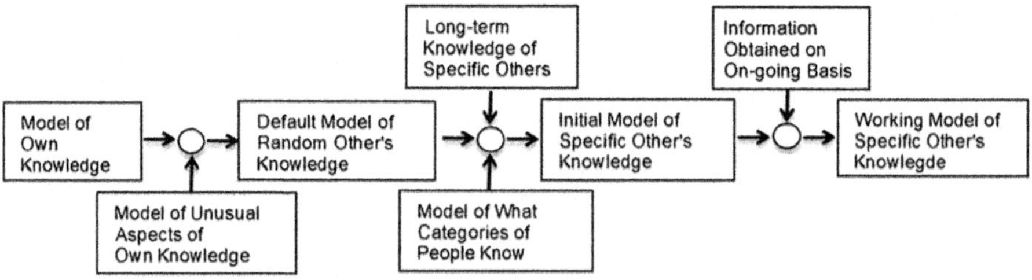

Fig. 4 A model on how people assume what other people know (From Nickerson, 1999)

by knowledge of *unusual aspects of own knowledge*, knowledge of what *categories of people* know, *long-term knowledge of specific others* and *information obtained* on *on-going basis*.

According to Nickerson, people strongly tend to project their own knowledge on other people because the initial anchor is insufficiently adjusted. Thus, the expert blind spot can be explained by the dominance of the anchor of the experts' knowledge. The process of estimating students' knowledge can be interpreted as such a decision under uncertainty (prone to the overestimation of students' performance), which underlies a broad variety of influencing factors.

Although Nickerson's model describes a rather intuitive process, it might give clues on the facets of knowledge that can help to adjust the anchor appropriately. For this purpose, components of MKT (Ball et al., 2008) can be related to the adjusting factors of Nickerson's model (Ostermann et al., 2017). These assignments can serve as a framework in order to generate hypotheses and designs for studies to investigate the influence of (adjusting) factors on teachers' judgment accuracy.

Earlier in this chapter, we have reflected on the role of specificity of judgments: Firstly, specific judgments, in the sense that teachers refer to a specific task, showed the strongest effects in the meta-analyses of Südkamp et al. (2012) and Hoge and Coladarci (1989); secondly, the group specificity of judgments played a role for the empirical investigation in order to avoid confounders like social biases. This distinction between group-specific and non-specific factors is also inherent in Nickerson's model. In Fig. 5, we present an adaption of the model which emphasizes the group-specificity of the adjusting factors, of which the non-specific ones can be related to components of MTK (Ostermann et al., 2017).

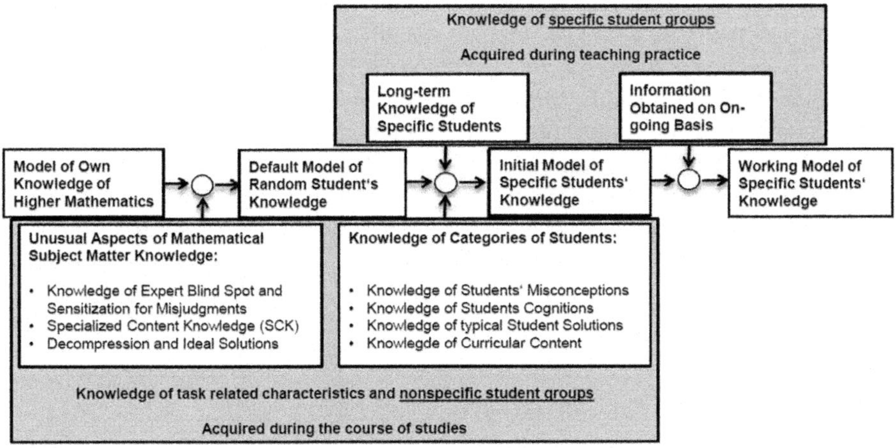

Fig. 5 A model on how teachers assume what students know (Adapted from Ostermann et al., 2017)

4 Conclusion

In this chapter, we discussed factors that influence the accuracy of teachers' diagnostic judgments. It seems that teachers' mathematical content knowledge (CK) plays a key role. It must be seen as an ambivalent facet: On the one hand, it is necessary for students' success (Anders, Kunter, Brunner, Krauss, & Baumert, 2010); on the other hand, research shows that diagnostic judgments can be biased by subject matter knowledge in accordance with the expert-blind-spot theory. This may happen if teachers project their own knowledge into students, which is a form of anchoring and inadequate adjustment (Nickerson, 1999; Ostermann et al., 2015). Nickerson's model of how people estimate other persons' knowledge can be used as a suitable framework to describe certain aspects of the process of diagnostic judgments. This model predicts the predominance of the estimators' own knowledge as an anchor and provides clues about specific (regarding specific persons) and non-specific factors that influence the process of estimation; for example, knowledge about what categories of people know. Thus, it could be shown that teachers are not necessarily influenced by a 'curse of expertise' (Hinds, 1999) but that they can be enabled to adjust their point of view by knowledge of students' perspectives (Ostermann et al., 2017). Knowledge of students' misconceptions (KCS) could be pointed out as one significant factor which improves the judgment accuracy (Hadjidemetriou & Williams, 2002a, 2002b; Leinhardt, 1990; Ostermann et al., 2017).

These findings were observed by investigating teachers' judgments on 'typical' but non-specific student groups. This approach is beneficial in order to avoid biasing anchoring effects, such as reference group effects (Dünnebier et al., 2009), that possibly emerge when estimating specific student groups. Thus, judgments of different teachers with reference to the same criterion can be compared with fewer biases in statistical analysis. Since that approach is rarely used up to now, further studies, which include both types of measurement, specific and non-specific student groups, would be beneficial in two respects: Firstly, the approach of estimating non-specific, representative groups could be further validated by comparing teachers' judgments in both components of Table 1. Secondly — but closely related to the first question — the significance and mutual influence of (group) specific and non-specific factors in Nickerson's model could be clarified. Particularly, the genesis of diagnostic abilities could be illuminated by think aloud interviews to shed light on how teachers integrate KCS in the process of estimation.

References

Anders, Y., Kunter, M., Brunner, M., Krauss, S., & Baumert, J. (2010). Diagnostische Fähigkeiten von Mathematiklehrkräften und ihre Auswirkungen auf die Leistungen ihrer Schülerinnen und Schüler. [Mathematics teachers' diagnostic skills and their impact on students' achievements]. *Psychologie in Erziehung und Unterricht*, *57*(3), 175–193.

Ball, D. L., Thames, M. H., & Phelps, G. (2008). Content knowledge for teaching what makes it special? *Journal of Teacher Education*, *59*(5), 389–407.

Baumert, J., Kunter, M., Blum, W., Brunner, M., Voss, T., Jordan, A., & Tsai, Y.-M. (2010). Teachers' mathematical knowledge, cognitive activation in the classroom, and student progress. *American Educational Research Journal, 47*(1), 133–180.

Betsch, C. (2004). Präferenz für intuition und deliberation (PID). *Zeitschrift für Differentielle und Diagnostische Psychologie, 25*(4), 179–197.

Chi, M. T., Siler, S. A., & Jeong, H. (2004). Can tutors monitor students' understanding accurately? *Cognition and Instruction, 22*(3), 363–387.

Dünnebier, K., Gräsel, C., & Krolak-Schwerdt, S. (2009). Urteilsverzerrungen in der schulischen Leistungsbeurteilung: Eine experimentelle Studie zu Ankereffekten. [Biases in teachers' assessments of student performance: An experimental study of anchoring effects.] *Zeitschrift für Pädagogische Psychologie, 23*(34), 187–195.

Evans, J. S. B. (2008). Dual-processing accounts of reasoning, judgment, and social cognition. *Annual Review of Psychology, 59*, 255–278.

Gigerenzer, G., & Gaissmaier, W. (2011). Heuristic decision making. *Annual Review of Psychology, 62*, 451–482.

Hadjidemetriou, C., & Williams, J. (2002a). Children's graphical conceptions. *Research in Mathematics Education, 4*(1), 69–87.

Hadjidemetriou, C., & Williams, J. (2002b). Teachers' pedagogical content knowledge: Graphs, from a cognitivist to a situated perspective. In A. D. Cockburn & E. Nardi (Eds.), *Proceedings of the 26th Conference of the International Group for the Psychology of Mathematics Education* (Vol. 3, pp. 57–64). Norwich, UK: PME.

Helmke, A., & Schrader, F.-W. (1987). Interactional effects of instructional quality and teacher judgement accuracy on achievement. *Teaching and Teacher Education, 3*(2), 91–98.

Herppich, S., Wittwer, J., Nückles, M., & Renkl, A. (2010). Do tutors' content knowledge and beliefs about learning influence their assessment of tutees' understanding. In S. Ohlsson & R. Catrambone (Eds.), *Proceedings of the 32th Annual Conference of the Cognitive Science Society* (pp. 314–319). New York, NY: Erlbaum.

Hinds, P. J. (1999). The curse of expertise: The effects of expertise and debiasing methods on prediction of novice performance. *Journal of Experimental Psychology: Applied, 5*(2), 205–221.

Hoge, R. D., & Coladarci, T. (1989). Teacher-based judgments of academic achievement: A review of literature. *Review of Educational Research, 59*(3), 297–313.

Kahneman, D. (2003). A perspective on judgment and choice: Mapping bounded rationality. *American Psychologist, 58*(9), 697.

Kaiser, J., Helm, F., Retelsdorf, J., Südkamp, A., & Möller, J. (2012). Zum Zusammenhang von Intelligenz und Urteilsgenauigkeit bei der Beurteilung von Schülerleistungen im Simulierten Klassenraum. [On the relation of intelligence and judgment accuracy in the process of assessing student achievement in the simulated classroom.] *Zeitschrift für Pädagogische Psychologie, 26*, 251–261.

Karst, K. (2012). *Kompetenzmodellierung des diagnostischen Urteils von Grundschullehrern.* Münster, German: Waxmann.

Karst, K., Schoreit, E., & Lipowsky, F. (2014). Diagnostische Kompetenzen von Mathematiklehrern und ihr Vorhersagewert für die Lernentwicklung von Grundschulkindern. *Zeitschrift für Pädagogische Psychologie, 28*(4), 237–248.

Kelley, C. M. (1999). Subjective experience as basis of "objective" judgments: Effects of past experience on judgments of difficulty. Effects of past experience on judgments of difficulty. In D. Gopher & A. Koriat (Hrsg.), Attention and performance (Bd. 17, S. 516–536). MIT Press.

Klug, J., Bruder, S., Kelava, A., Spiel, C., & Schmitz, B. (2013). Diagnostic competence of teachers: A process model that accounts for diagnosing learning behavior tested by means of a case scenario. *Teaching and Teacher Education, 30*, 38–46.

Leuders, T., Bruder, R., Kroehne, U., Naccarella, D., Nitsch, R., Henning-Kahmann, J., & Wirtz, M. (2017). Development, validation, and application of a competence model for mathematical problem solving by using and translating representations of functions. In *Competence Assessment in Education* (pp. 389–406). Springer International Publishing.

Leuders, T., Dörfler, T., Leuders, J., & Philipp, K. (2018). Diagnostic competence of mathematics teachers – unpacking a complex construct. In T. Leuders, K. Philipp, & J. Leuders (Eds.), Diagnostic competence of mathematics teachers. Cham: Springer.

Lin, J. W., & Chiu, M. H. (2010). The mismatch between students' mental models of acids/bases and their sources and their teacher's anticipations thereof. *International Journal of Science Education*, *32*(12), 1617–1646.

Morris, A. K., Hiebert, J., & Spitzer, S. M. (2009). Mathematical knowledge for teaching in planning and evaluating instruction: What can preservice teachers learn? *Journal for Research in Mathematics Education*, *40*, 491–529.

Nathan, M. J., & Koedinger, K. R. (2000). An investigation of teachers' beliefs of students' algebra development. *Cognition and Instruction*, *18*(2), 209–237.

Nathan, M. J., & Petrosino, A. (2003). Expert blind spot among preservice teachers. *American Educational Research Journal*, *40*(4), 905–928.

Nickerson, R. S. (1999). How we know—And sometimes misjudge—What others know: Imputing one's own knowledge to others. *Psychological Bulletin*, *125*(6), 737–795.

Nickerson, R. S. (2001). The projective way of knowing: A useful heuristic that sometimes misleads. *Current Directions in Psychological Science*, *10*(5), 168–172.

Ostermann, A., Leuders, T., & Nückles, M. (2015). Wissen, was Schülerinnen und Schülern schwer fällt. Welche Faktoren beeinflussen die Schwierigkeitseinschätzung von Mathematikaufgaben? [Knowing what students know. Which factors influence teachers' estimation of task difficulty?]. *Journal für Mathematik-Didaktik*, *36*(1), 45–76.

Ostermann, A., Leuders, T., Philipp, K. (2017). Fachbezogene diagnostische Kompetenzen von Lehrkräften – Von Verfahren der Erfassung zu kognitiven Modellen zur Erklärung. In Leuders, T., Nückles, M., Mikelskis-Seifert, S. & Philipp, K. (Eds.), Pädagogische Professionalität in Mathematik und Naturwissenschaften. Springer

Plessner, H., & Czenna, S. (2008). The benefits of intuition. In H. Plessner, C. Betsch, & T. Betsch (Eds.), *Intuition in judgment and decision making* (pp. 251–265). Mahwah, NJ: Lawrence Erlbaum Associates Publishers.

Plessner, H., Freiberger, V., Kurle, A., & Ochs, K. (2006, May). *Besser früh und spontan als spät und analytisch: Die Genauigkeit von Expertenvorhersagen für den FIFA Confed-erations Cup 2005* [Better early and spontaneous than late and analytic: The accuracy of expert predictions concerning the FIFA Confederations Cup 2005]. Paper presented at the 36th Meeting of the Arbeitsgemeinschaft Sportpsychologie in Münster, Germany.

Rikers, R. M., Schmidt, H. G., & Boshuizen, H. P. (2000). Knowledge encapsulation and the intermediate effect. *Contemporary Educational Psychology*, *25*(2), 150–166.

Schrader, F.-W. (2006): Diagnostische Kompetenz von Eltern und Lehrern (3. Aufl.). In D. H. Rost (Hrsg.), *Handwörterbuch Pädagogische Psychologie* (3. Aufl., S. 95–100). Weinheim, Germany: Beltz.

Selter, C. (1995). Zur Fiktivität der Stunde Null im arithmetischen Anfangsunterricht. *Mathematische Unterrichtspraxis*, *16*(2), 11–19.

Spinath, B. (2005). Akkuratheit der Einschätzung von Schülermerkmalen durch Lehrer und das Konstrukt der diagnostischen Kompetenz. [Accuracy of teacher judgments on student characteristics and the construct of diagnostic competence]. *Zeitschrift für Pädagogische Psychologie*, *19*(1), 85–95.

Südkamp, A., Kaiser, J., & Möller, J. (2012). Accuracy of teachers' judgments of students' academic achievement: A meta-analysis. *Journal of Educational Psychology*, *104*(3), 743–762.

Südkamp, A., & Möller, J. (2009). Referenzgruppeneffekte im Simulierten Klassenraum: direkte und indirekte Einschätzungen von Schülerleistungen. [Reference-group-effects in a simulated classroom: Direct and indirect judgments.] *Zeitschrift für Pädagogische Psychologie*, *23*(34), 161–174.

Tversky, A., & Kahneman, D. (1974). Judgment under uncertainty: Heuristics and biases. *Science*, *185*(4157), 1124–1131.

Vollrath, H.-J. (1989). Funktionales Denken. *Journal für Mathematik-Didaktik*, *10*(1), 3–37.

Wilson, T. D., & Schooler, J. W. (1991). Thinking too much: Introspection can reduce the quality of preferences and decisions. *Journal of Personality and Social Psychology*, *60*(2), 181.

Diagnostic Competences of Mathematics Teachers with a View to Processes and Knowledge Resources

Kathleen Philipp

Diagnostic competences of teachers are considered important for students' learning success. However, there is little empirical evidence about cognitive processes of teachers in (informal) diagnostic situations and the knowledge they use in such situations. The concern of the reported study is to extend this state of research from a domain-specific point of view. For this purpose, processes and knowledge resources mathematics teachers apply in informal diagnostic situations (evaluating tasks and students' solutions) are identified empirically and described theoretically. The findings show that the teachers in the study proceed predominantly in a systematic way and use a variety of different types of mathematical knowledge.

1 Introduction: Diagnostic Competence

In everyday teaching we find various kinds of diagnostic situations. They can be characterized with regard to their purpose and position in a learning process (e.g., Ingenkamp & Lissmann, 2008; Wiliam, 2007).

At the beginning of a learning process an *initial assessment* can yield information about students' previous knowledge and preconditions for planning lessons. During a learning process, *formative assessment* can be used to support students individually or adapt lessons. *Summative assessment* at the end of a learning process can be used for assessing learning results, grading or evaluating instruction (Fig. 1).

Another way to characterize diagnostic situations is their level of formality. In addition to *formal assessment* (e.g., by standardized diagnostic tests), *informal*

K. Philipp (✉)
School of Education, University of Applied Sciences and Arts Northwestern Switzerland, Basel, Switzerland
e-mail: kathleen.philipp@fhnw.ch

T. Leuders et al. (eds.), *Diagnostic Competence of Mathematics Teachers*, Mathematics Teacher Education 11, DOI 10.1007/978-3-319-66327-2_6

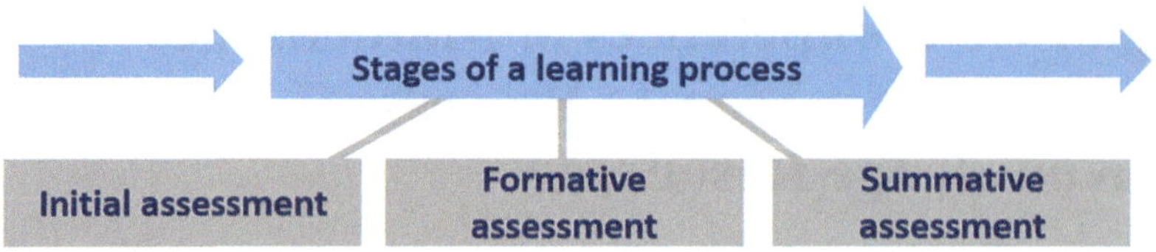

Fig. 1 Diagnostic situations in a learning process

assessment (e.g., by observation) plays an important role in the classroom, e.g., by selecting appropriate tasks or reacting to students' mistakes.

Within diagnostic situations one can also distinguish a *preactional*, an *actional* and a *postactional phase*, each phase characterized by different types of diagnostic judgments (Klug, Bruder, Kelava, Spiel, & Schmitz, 2013): In the preactional phase, teachers plan diagnostic actions by selecting methods, aims, criteria, and so on. In the actional phase, they collect data, interpret and draw conclusions and in the postactional phase they enact instructional measures (Klug et al., 2013). This tripartite model allows a broader understanding of diagnostic situations as it includes diagnostic activities of teachers before and after the actual diagnostic judgment.

Diagnostic competence can be defined as the individuals' dispositions that are necessary to manage diagnostic situations successfully, and it can be seen as one of the key competences for teaching (Helmke, Hosenfeld, & Schrader, 2004). There is a broad agreement that diagnostic competence is essential for the quality of teaching; for example, when teachers have to select adequate measures such as modifying the difficulty of a task or when they have to adapt educational methods (Helmke et al., 2004; Anders, Kunter, Brunner, Krauss, & Baumert, 2010; Schwarz, Wissmach, & Kaiser, 2008). However, we find hints that diagnostic competence of teachers is not sufficiently developed (e.g., Krauss & Brunner, 2011).

In recent studies diagnostic competence often is described as the ability of a person to judge people accurately and is measured as the precision of certain judgments (Schrader, 2011). In these judgments teachers have to estimate the level, the variance and the ranking order of students' attributes or of tasks. Such a numerical precision can be regarded as an indicator for diagnostic competence. However, this approach does not provide any understanding of the way teachers generate diagnostic judgments and their underlying cognitive processes (for a detailed discussion see Leuders, Dörfler, Leuders & Philipp 2018 – chapter "Diagnostic Competence of Mathematics Teachers: Unpacking a Complex Construct" in this book).

In order to support the development of diagnostic competence of teachers, it is important to understand the processes and identify the knowledge teachers apply during these processes (Barth & Henninger, 2012). Furthermore, it would be a matter of particular interest to clarify the domain-specificity (or even topic-specificity) of diagnostic competence and its structure. We find indications that diagnostic competence should not be understood as a general ability, but rather as a construct composed of multiple partial competences (Spinath, 2005). In conclusion, we still need further research to develop a better theoretical understanding of processes that underlie teachers' diagnostic judgments (see also the concept of diagnostic thinking in Leuders et al. 2018 in this book).

Accordingly, the analysis in this chapter deals with the following questions: (1) How do teachers arrive at their diagnostic judgments, and (2) what kind of knowledge do they need in diagnostic processes? The study reported here focuses on informal diagnostic situations in mathematics, such as evaluating tasks and evaluating students' solutions. These situations can be described as the actional phase in which teachers have to gather and interpret information about tasks and students' knowledge. Both diagnostic situations can occur at every position in a learning process.

In the following section the theoretical framework of the study is described, particularly recurring on models that may help understand diagnostic processes and knowledge resources of teachers in these diagnostic situations.

2 Theoretical Framework

In order to describe diagnostic competence(s) as judgment processes within a theoretical framework, it is helpful to take a closer look at suitable models that focus on cognitive processes and their underlying knowledge resources, that is, which tap diagnostic thinking (within the general model of diagnostic competence as a continuum, cf. chapter "Diagnostic Competence of Mathematics Teachers: Unpacking a Complex Construct" in this book). In this section three theoretical approaches were taken into consideration to identify the cognitive processes and knowledge resources in diagnostic situations. The models vary in their domain-specificity and therefore provide an insight from different perspectives on diagnostic activities. First, the focus is on processes in estimating other people's knowledge in general and then mathematical knowledge resources for teachers were delineated. The third approach gives a first insight into the interaction of processes and knowledge resources in diagnostic situations.

2.1 Diagnostic Processes as an Alternation of Anchoring and Adjustment

Diagnostic situations in the classroom require the evaluation of the current level of students' knowledge by the teacher. With the exception of the use of standardized measuring instruments, such diagnostic situations can be related to situations described in the field of research on expertise: The rating of other people's knowledge, especially the rating of the knowledge of novices by experts (Ostermann, Leuders, & Philipp, 2017; Philipp & Leuders, 2014).

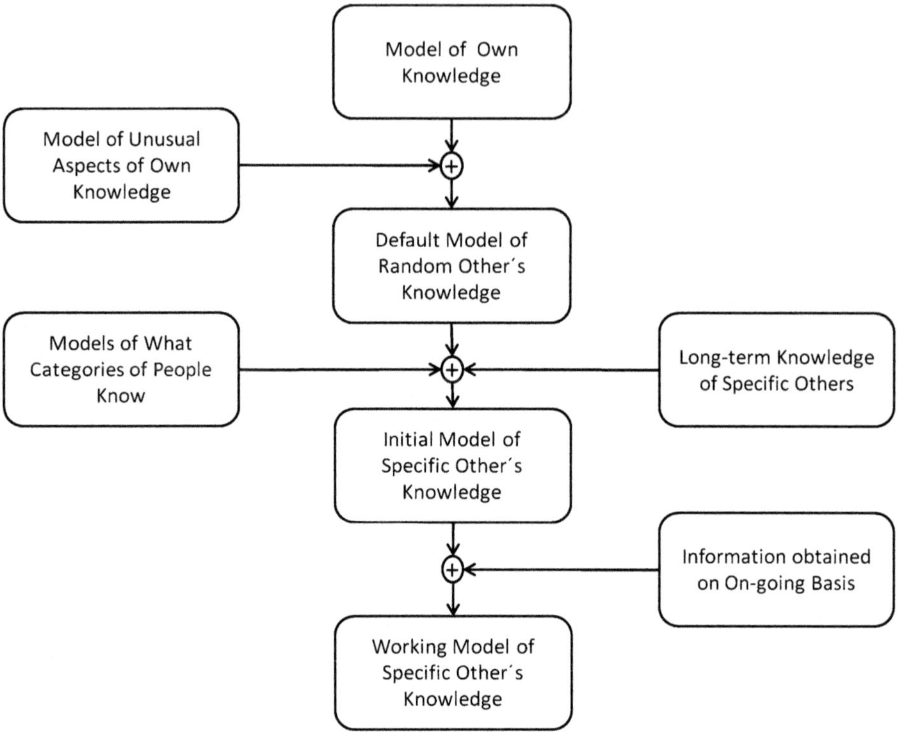

Fig. 2 Model of rating other people's knowledge (From Nickerson, 1999, p. 740)

Nickerson (1999) proposes a model (see Fig. 2) to describe this process in three steps: (1) The expert uses his or her knowledge as a basis (*model of own knowledge*) and keeps the exclusivity of his or her knowledge in mind (e.g., specific knowledge of teachers). This leads to a *default model of a random other's knowledge*. This default model represents a kind of common knowledge of any person (or a group of persons) and builds the foundation for further, more specific, models. (2) The consideration of information such as the affiliation to a specific group of people (e.g., class level) and information from former experience are used to modify the model and result in an *initial model of a specific other's knowledge*. This model construction is not necessarily a conscious process. (3) The process of rating other people's knowledge continues in gaining information about a specific person (e.g., in conversation) and leads to a *working model of a specific other's knowledge*. So the development of a model of other people's knowledge is a permanent refinement and update. The whole process can be characterized as a heuristic of "anchoring an adjustment" (Tversky & Kahnemann, 1974), in which the own knowledge of the expert as starting point plays an important role. In considering aspects of knowledge continuously, this heuristic leads to different steps of a model of other peoples' knowledge. The different steps then can be understood as anchors in building the starting point for further adjustment.

In the classroom teachers frequently have to assess the knowledge of students, groups of students or a whole class. The continuous process of modifying and updating such assessments (regarded as anchoring and adjustment process) may occur constantly during a lesson and enables adaptive teaching. Nickerson's model, when applied on pedagogical situations, can be helpful to understand diagnostic processes; especially, since it helps to explain biases, which frequently occur – generally and in the pedagogical context:

- *False-consensus effect*: Own opinions and attitudes are considered to be opinions and attitudes of the majority of people (Brown, 1982).
- *Egocentric bias*: The attribution of general knowledge to other people is strongly connected to own general knowledge (Nickerson, 1999).
- *Curse of expertise*: With increasing expertise challenges for novices are underestimated (e.g., Camerer, Loewenstein, & Weber, 1989).
- *Illusion of simplicity*: Experts misjudge topics as simple because they are familiar with it (Kelley, 1999).
- *Hindsight bias*: It is difficult to reconstruct the own state of previous knowledge (Fischhoff, 1975).

This shows that tendencies of overestimating knowledge of others are frequent. Such tendencies can be explained by insufficient adaptation in Nickerson's model or deficient awareness of the exclusivity of one's own knowledge. In pedagogical situations this can have severe consequences: The discrepancy of mathematical expert knowledge of teachers (after several years of teacher education and teaching practice) and the knowledge of students is enormous. Furthermore, it is difficult to adopt a novice's perspective: "Every beginning instructor discovers sooner or later that his first lectures were incomprehensible because he was talking to himself, so to say, mindful only of his own point of view. He realizes only gradually and with difficulty that it is not easy to place oneself in the shoes of students who do not know what he knows about the subject matter of his course" (Piaget, 1962, p. 5, in Nickerson, 1999, p. 747).

A transfer to pedagogical situations seems plausible – the assessment of students' knowledge guides the teachers' actions in the classroom. On the other hand, in the classroom there is also an important difference to the kind of situations Nickerson refers to: Teachers usually do have prior knowledge about the knowledge of their students. Even if they teach a new class, they do not only refer to their own knowledge to assess students' knowledge, but also consider general information such as age and class level. It can be assumed that such models of knowledge are increasingly used as foundation with growing teaching experience.

The typical misjudgments referred to above also exist in pedagogical situations: The *curse of expertise* often appears in context of achievement tests in which teachers falsely interpret correct results as confirmation of students' knowledge and thereby overestimate the knowledge of students. However, the reverse also occurs: Teachers sometimes underestimate the knowledge of students, supposing that they had no learning opportunity outside the classroom (Clarke et al., 2002; Selter, 1995). In Nickerson's model this can be interpreted as attributing a lack of knowledge, for example, to first-graders. In other situations, it is apparent that teachers use the knowledge of good students as standard for the whole class (Schrader & Helmke, 1987). The awareness of such biases can also be used in a productive way; for example, in teaching mathematics the *hindsight bias* can be used in designing teaching situations bearing in mind a "preview-perspective" of students that are not familiar with a content and a "review-perspective" of the teacher (Ruf & Gallin, 2005). These examples show that Nickerson's model can be used to understand frequent tendencies of over- or underestimating students' knowledge in pedagogical context and it also highlights the significance of own knowledge.

With regard to diagnostic situations in the classroom Nickerson's model is helpful for understanding the process of generating judgments and it also helps to gain insight into resources people use when assessing other people's knowledge. Besides, typical misjudgment tendencies can be interpreted within the model. The model is very general and therefore can be transferred to a variety of situations. This generality also leads to limitations: It does not contribute to our understanding of domain-specific processes.

2.2 Diagnostic Competences as Facet of Professional Knowledge for Teaching Mathematics

In order to analyze diagnostic processes, it is not sufficient to describe the *processes* of generating knowledge on students, as in section 2.1. One should also be aware of different *types* of knowledge that are relevant in the process.

Several attempts to describe domain-specific types of teacher knowledge can be traced back to the framework of Shulman who categorized teacher knowledge and introduced pedagogical content knowledge as "the category most likely to distinguish the understanding of the content specialist from the pedagogue" (Shulman, 1987, p. 8). Pedagogical content knowledge includes knowledge about typical difficulties of students and their pre- or misconceptions (Shulman, 1986). This category already hints at the concept of diagnostic competence.

For the domain of mathematics, Shulmans' categories were refined and substantiated in various studies by Ball and colleagues (Ball, Lubienski, & Mewborn, 2001; Ball, Thames, & Phelbs, 2008; Hill, Ball, & Schilling, 2008), for example. Their detailed analyses of teaching practice led to a categorization of mathematical knowledge applied in teaching (Ball et al., 2008, Fig. 3). In their job analysis, they considered teaching situations in the classroom and activities connected to teaching such as lesson planning, managing homework or evaluating students' work.

When considering competences required in diagnostic situations, several of these knowledge domains can be regarded as relevant:

- First of all, in diagnostic situations it is essential to evaluate the mathematical correctness of a solution. The knowledge needed for this pertains to the domain

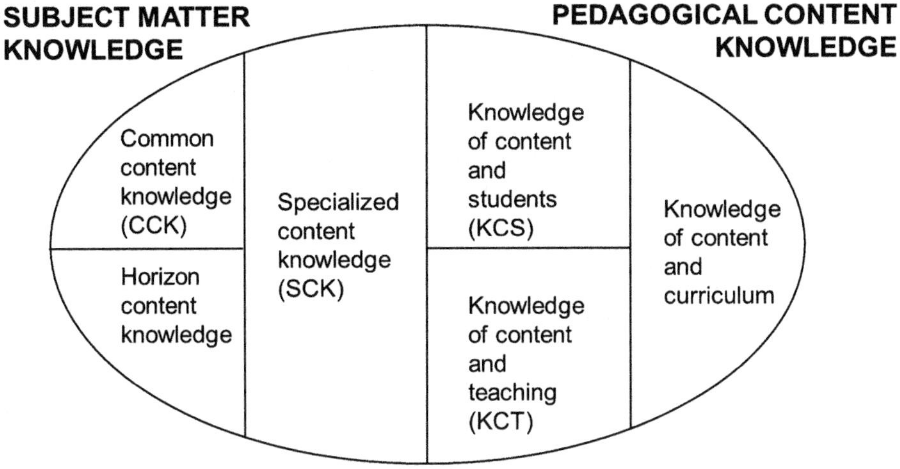

Fig. 3 Domains of mathematical knowledge for teaching (From Ball et al., 2008, p. 403)

of *common content knowledge (CCK)*. In diagnostic situations it is often necessary to decide very quickly if a solution is correct or a solution attempt is adequate.

- In contrast to common content knowledge *specialized content knowledge (SCK)* is considered as a kind of subject matter knowledge that is required only for teaching. It goes beyond the understanding of mathematical contents people in other fields need because the teaching of mathematical contents requires a deeper understanding to make it learnable for students. In diagnostic situations it is used, for example, to modify the difficulty of tasks or find patterns in students' errors. Note that this kind of knowledge draws on mathematical knowledge and does not require knowledge about students.
- The domain *knowledge of content and students (KCS)* is characterized by a close connection of mathematical knowledge and knowledge about students and may be the most important for diagnostic situations. For example, it is crucial to know typical errors or students' (mis-)conceptions within mathematical topics to follow their way of thinking.

The theoretical approach of Ball and colleagues to identify and substantiate several domains of mathematical knowledge for teaching is also useful to capture knowledge for diagnostic activities of teachers; particularly, because the mentioned knowledge domains are closely connected to typical activities of mathematics teachers. Although the model does not describe diagnostic processes directly, it can be helpful in understanding different kinds of mathematical knowledge needed in diagnostic situations and contributes to clarify knowledge resources which teachers use in such situations.

2.3 Diagnostic Processes as Unpacking Learning Goals

Morris, Hiebert, and Spitzer (2009) construct a theoretical model for teacher thinking which applies to a specific diagnostic situation. They consider the ability to "unpack" the sub-goals of a task as important for planning and evaluating students' learning processes. They describe this ability as a type of mathematical knowledge that is special for teaching quite similar to the SCK by Ball et al. (2008), as described above. As an example, Morris et al. decompose the learning goal "students will understand how to add fractions and will understand the concepts underlying this operation" into six sub-goals that are necessary to attain the learning goal (Morris et al., 2009, p. 499):

1. A quantity is identified as the quantity "one."
2. We obtain units of size $1/n$ by partitioning the "one" into n equal parts.
3. The numerator is the number of units of size $1/n$.
4. The addends must both be expressed in terms of the same-sized unit.
5. The addends must be joined.
6. The sum must be expressed in terms of a unit of size $1/n$.

This kind of analysis is a mathematical one and does not require any knowledge about individual student solutions. The authors emphasize that other decompositions of the learning goal are also possible. The identified sub-goals can be interpreted as subconcepts of students, and used to clarify students' errors.

In their study preservice teachers had to complete four written tasks. They had to (1) anticipate an ideal student response, (2) evaluate a student's incorrect response, (3) evaluate a student's correct response, and (4) analyze a classroom lesson (Morris et al., 2009). Especially, the activities in tasks (2) and (3) can be understood as diagnostic activities. One result of the study is that the participants do not decompose learning goals spontaneously but they could be supported to do so when referring to subconcepts in the tasks explicitly.

Such a decomposition of learning goals provides a better understanding of the students' failure and helps to localize it. However, the authors' example is a very specific one, since it is mostly based on an analysis of procedures. Typical misconceptions of students such as the idea of "division makes the result smaller", which is a learning obstacle when students go from natural numbers to fractions (e.g., Swan, 2001, p. 154) are not taken into consideration. As we can see by this example, it is not only the mathematical structure of a task, but also the structure of the learners' knowledge that leads to errors in students' solutions. Nevertheless, the decomposition of learning goals can be considered as an important facet of diagnostic competence. With regard to the identification of diagnostic processes and knowledge resources, the approach of Morris et al. comprises two aspects: The process of decomposing learning goals requires *specialized content knowledge* and knowledge about typical pre- or misconceptions of students (*knowledge of content and students*) (Ball et al., 2008).

The models discussed in the preceding sections contribute valuable theoretical ideas and empirical findings regarding diagnostic activities of teachers. Nickerson (1999) provides a very general model of an expert-novice-situation. This situation is similar to a diagnostic situation in the classroom although the author himself does not transfer the model into pedagogical context. Knowledge about frequent tendencies of under- or overestimating knowledge in the process of assessing other people's knowledge can be helpful for a deeper understanding of fundamental processes concerning own knowledge. Ball et al. (2008) suggest different domains of mathematical knowledge by analyzing teaching situations which are also helpful for identifying knowledge resources required in diagnostic situations. Morris et al. (2009) outline a very specific process of decomposing mathematical knowledge which can be considered relevant for diagnostic situations, for example, for analyzing a task. Thus, all these models, though of very different scope, can be useful for a deeper understanding of diagnostic processes and the knowledge resources needed. However, they have their limitations. Nickerson (1999) describes cognitive processes in a judgment process but is not clear to what extent such processes are relevant in the pedagogical context, especially, in diagnostic situations of mathematics teachers. On the other hand, the model of decomposing learning goals (Morris et al., 2009) seems to be too specific to describe diagnostic processes extensively. The ability to decompose can be understood as an important part of diagnostic competences but it does not consider students' misconceptions. Both models take knowledge resources needed in the processes into account only partially. The domains of mathematical knowledge (Ball et al., 2008) provide a framework that differentiates mathematics teachers' knowledge but it is not focused on diagnostic processes.

Taken together these models can be helpful in the understanding of diagnostic processes and knowledge resources and their interaction in diagnostic situations. However, an empirical analysis is needed that accounts for the specificity of diagnostic situations on the one hand and the variety of phenomena encountered when analyzing teacher behavior on the other.

3 Investigating Processes and Knowledge Resources in Diagnostic Situations

The study reported here focuses on informal diagnostic situations in which mathematics teachers (1) judge tasks and (2) evaluate students' solutions. Such diagnostic situations occur rather often: Teachers have to select and use tasks and have to react to students' solutions and mistakes spontaneously. To investigate the diagnostic competence of mathematics teachers, it seems essential to study real processes and the knowledge resources needed in such diagnostic situations. So, the research questions are:

1. What kind of *processes* do teachers show in their diagnostic judgments?
2. What kind of *knowledge* do teachers rely on during these processes?

Investigating these questions can result in a deeper understanding of diagnostic processes on the one hand and a clarification of constituent parts of diagnostic competence in mathematics on the other. Another (long-term) objective connected with these questions is to derive consequences for teacher education (Philipp & Leuders, 2014).

3.1 Design

In order to gain knowledge on diagnostic processes and on knowledge resources, two-phased interviews were conducted. In the first phase the teachers first had to evaluate two tasks. Then three students' solutions were presented successively to each task to initiate a diagnostic process. In the second phase the participants had to reflect on their own process in describing or explaining it. This procedure was expected to be beneficial for catching most of the relevant diagnostic processes and knowledge resources by triggering the participants with the tasks and students' solutions and afterwards having them describe their own processes and give some additional statements. The reasoning in both phases was captured by means of think-aloud interviews (Ericsson & Simon, 1993).

The participants were six experienced mathematics teachers, three of them additionally experienced in mathematics teacher education. That way it was possible to draw on practical experience and theoretical knowledge similarly. The aim was to find a broad variety of different processes and knowledge resources used in diagnostic situations. Think-aloud-protocols of the interviews provide the data for the analyses which were a total of 12 evaluations of tasks and 36 evaluations of students' solutions. The tasks were chosen from the topic of fractions due to the fact that broad systematic knowledge about typical students (mis-)conceptions in this field is available.

Figure 4 shows the two tasks and the three students' solutions to each task used in the interviews. The students' solutions were selected with regard to typical mistakes and misconceptions occurring frequently. The tasks and the interview guidelines were developed in a pilot study. In the first phase of the interview the participants were asked the following: "Please evaluate the task. How can you use it in the classroom?", and then, to evaluate students' solutions: "Please evaluate the students' solution. Which conclusions do you draw?" In the second phase the participants reflected their own processes by answering the questions: "How did you come to your evaluation? Please describe and give reasons for your procedure. What kind of knowledge did you use?"

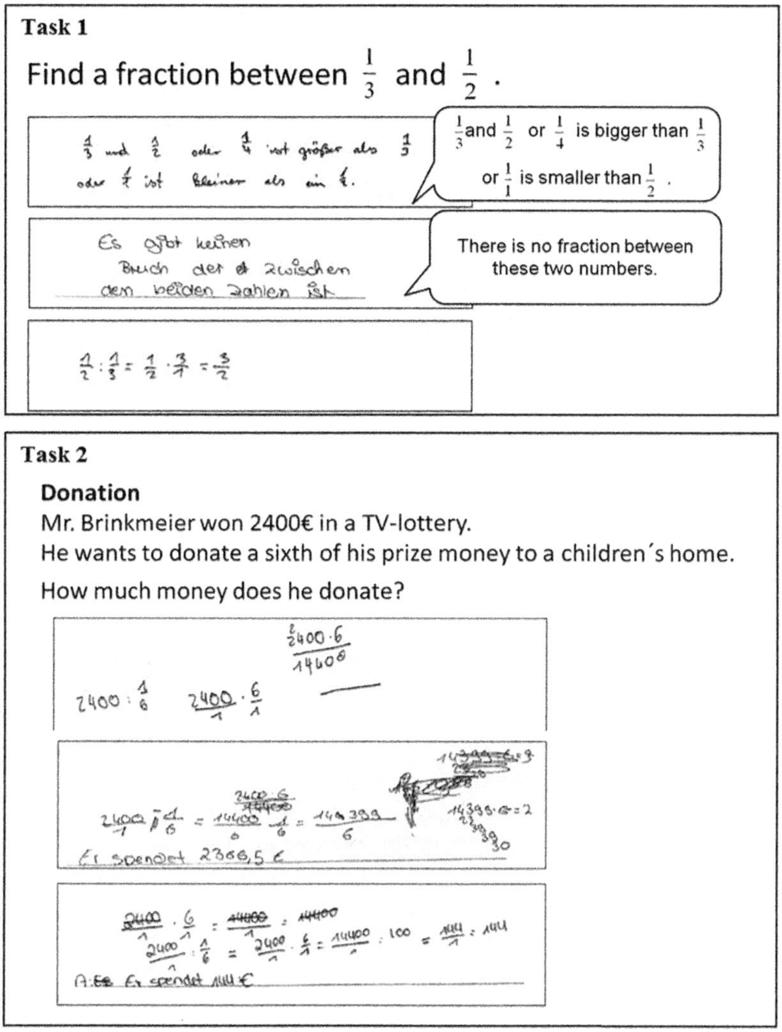

Fig. 4 Tasks and students' solutions (From: Wartha, 2007)

3.2 *Data Analysis*

For the analysis of the data interpretative content analysis, techniques were used (Mayring, 1983). The objectives were to build a theory of processes in diagnostic situations (research question 1) and generate hypotheses for further research in this area. To answer the second research question, the same data was analyzed with a focus on different types of knowledge that influence diagnostic processes. In order

Table 1 Theoretical categories for processes and knowledge resources

Processes	Knowledge resources[a]	Literature references
Using an anchor	Own knowledge	Nickerson (1999)
Adjust a model by using "new" information	knowledge of categories of people	
Decompose a learning goal	Mathematical procedures	Morris et al. (2009)
	Common content knowledge	Ball et al. (2008)
	Specialized content knowledge	
	Knowledge of content and students	

[a]Because of the design of the study (also participants with experience in teacher education and therefore not having own students) it didn't make sense to use a category like "use prior information about the student" which could also be derived from Nickerson's model. For a possible further study with real students of the participants, it could be an important category to consider

to answer both research questions, the following steps were carried out: First theoretical categories were built based on the models mentioned before for both processes and knowledge resources. Initial point for the analysis built the categories in table 1.

Based on the think-aloud-protocols code descriptions, examples and coding rules were defined. In the next step new categories were continually supplemented and specified by identifying further processes or knowledge resources. Thus, the development of deductive and inductive categories was necessary. The analyses were carried out using the qualitative data analysis software, MAXQDA.

To capture diagnostic processes, the think-aloud-protocols in the two diagnostic situations, the evaluation of tasks and the evaluation of students' solutions (first phase of the interview) were analyzed along with the reflection of the own processes (second phase of the interview). These different approaches provided an insight into a broad variety of diagnostic processes. Table 2 shows the identified processes in detail.

The same data was analyzed again to answer the question which knowledge teachers rely on in such diagnostic situations (see Table 3). For example, the teacher statement "So that 1/4 is larger than 1/3. So the typical error, that, that it turns around, when it is in the denominator. With larger and smaller." can illustrate the proceeding in the analysis. With a focus on processes, this excerpt can be interpreted as *identifying deficits*. The same statement also gives an indication of knowledge resources the teacher draws on: a *typical mistake*. So, this example strikingly shows the interaction of processes and knowledge resources, although this is not the case in every statement in equal measure.

The generation of all codes, both processes and knowledge resources, were discussed several times in a group of researchers in mathematics education. In addition, they were used at two different points in time for the same data to assure the consistency of the assignments to categories.

Table 2 Diagnostic processes

Code	Definition	Representative teacher statement
Solution approach	Design a solution for a given task.	"[…] you can solve it by division."
Identify prerequisites/ barriers	State needs of a task or possible barriers for students.	"Students need a clear idea that you can expand and reduce a fraction."
Follow students solution	Reconstruct the student's approach.	"Yes, basically, he divided by 1/6."
Identify strengths	Discover and classify skills.	"[…] this is great. He writes down the number 2400 as fraction."
Identify deficits	Discover and classify errors.	"So that 1/4 is larger than 1/3. So the typical error, that, that it turns around, when it is in the denominator. With larger and smaller."
Error hypotheses	Give possible reason(s) for mistakes.	"[…] typical misconception that occurs when trying to transfer knowledge about natural numbers."
Measures to test hypotheses	State possibilities to verify an error hypotheses.	"[…] ask him to depict it by a picture."
Taking students perspective	Follow the students' argumentation from their point of view.	"He thought the numerator is equal (..) the denominator is not equal and between 3 and 2 I don't know a number."
Analysing step by step	Decompose a task or a student's solution.	"And then it goes on […] and now it gets interesting […]"
Comparing with	Compare the students solution with an own solution approach, instruction, mental models or familiar mistakes.	"So, I have in mind, how I teach fractions."

Table 3 Knowledge resources used in diagnostic situations

Code	Definition	Representative teacher statement
Correctness	Refer to mathematical background.	"Rule was recalled: multiply with the reciprocal value."
Mental models	Refer to topic-specific (basic) concepts, often with reference to literature.	"[…] for example basic concepts of fractions, the different kinds […]."
Different representations	Use different representations for a mathematical content.	"It becomes easier if you also use a good visualization."
Multiple approaches	Create several solution approaches to a task.	"[…] as teacher you need different ways to solve tasks, yes, to help students with difficulties."
Typical mistakes	Refer to common topic-specific errors.	"This is what you expect. You know that over the years."
Typical misconceptions	Refer to common topic-specific misconceptions.	"Some typical misconceptions with fractions you always have in mind."
Student strategies	Refer to typical students strategies (independent from leading to a correct or a wrong solution).	"And here you often find this solution, a 'point 5'-solution, […] so you have 2,5/6."
Diagnostic methods	Use methods to find reasons for errors or misconceptions.	"So, what I like doing, is 'finding the error' with the students, […] the students are the diagnosticians […] for example 'fraction-detective'."

In a further step of data analysis, relationships between the identified codes were investigated. Such relations were frequently occurring sequences of two (or more) codes or their simultaneous appearance. Also relationships between processes and knowledge resources were included in the analysis in order to achieve a more comprehensive view. The main result of this step of the analysis is a model of diagnostic processes presented in section 3.3.

3.3 Results

The intention of this section is not only to report the results of the study but also to point out their relation to the theoretical approaches mentioned above. By considering the relations of the identified codes, the diagnostic processes (first research question) can for the most part be interpreted as sequence of steps. It should be noted that not all of the participants showed all processes and there are also different qualities of individual processes observable in every step. This may be due to the use of different knowledge resources in the steps.

Figure 5 shows an idealized model of diagnostic processes and their relations: When the participants have to evaluate a task, the starting point often is an own solution approach. Then the participants identified prerequisites required by the task, and also potential barriers. In order to evaluate a student's solution, it is necessary to follow the student's argumentation. Thereby, strengths as well as deficits in the student's approaches can be identified. Possible reasons for mistakes or misconceptions can be expressed by error hypotheses. In addition, measures to investigate if an error hypothesis is correct can be proposed.[1]

Apart from these steps in a diagnostic situation, processes were found that typically comprise more than one step. These processes can be interpreted as strategies in diagnostic situations (Table 3). When the participants made an own solution approach, identified prerequisites and barriers of a task or followed the students' solution (first three steps), it became clear that they took the student's perspective. For example, they tried to adopt the thinking of an imaginary student in analyzing a task or tried to reconstruct the thinking of a particular student in analyzing the given students' solutions. A very common strategy is to decompose tasks or students' solutions and analyze them step by step. In order to identify strengths and deficits students' solutions are compared; for example, with own solution approaches or typical mistakes.

With reference to the theoretical models, the relevance of the teachers' own knowledge becomes apparent. It constitutes the fundament for diagnostic processes, as can be seen in the own solution approach, for example. This is also essential in

[1] This sequence can be influenced by the design of the interview situation: Tasks had to be evaluated first and the students' solutions afterwards. However, this situation is very close to typical situations in the classroom where the teachers first think about tasks when selecting them and then have to deal with the students' solutions to the tasks. Thus, it seems to be a "natural" procedure.

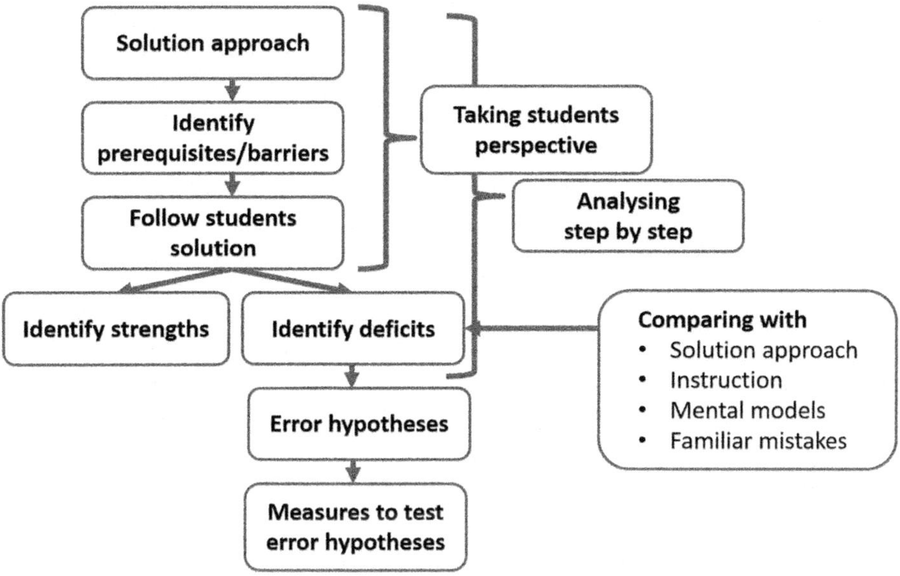

Fig. 5 Model of diagnostic processes

the model Nickerson (1999) proposes as starting point of the process of assessing other people's knowledge. In Nickerson's approach the model of other people's knowledge then is refined and adjusted. This is similar to the process of taking a students' perspective in diagnostic situations; for example, when a teacher considers skills or typical strategies of a 6th grader, and can be seen as adaptation of an initial model of other peoples' knowledge. When tasks or solutions are analyzed step by step, this can be interpreted as a kind of decomposition described by Morris and his colleagues (2009). Although it is not the specific decomposing of a learning goal, the decomposing of the prerequisites of a task, for example, so the strategy seems to be similar. This strategy also occurs in the evaluation of students' solutions by analyzing their way of thinking stepwise. In the same manner decomposing is not restricted to learning goals. Thus, the understanding of decomposing here is an extended one.

Concerning the second research question regarding the knowledge resources teachers rely on in diagnostic situations, the codes can be related to different types of (mathematical) knowledge:

- Teachers refer to the *correctness* of a solution when they have to decide which approaches are mathematically productive. This is mathematical content knowledge.
- The use of *mental models* and *different representations* is helpful when teachers evaluate tasks, for example. Having *multiple approaches* available can be useful to identify prerequisites or barriers of a task. This can be described as knowledge only teachers need.

- Topic-specific knowledge about *typical mistakes*, *typical misconceptions* or *student strategies* is a type of mathematical knowledge that includes knowledge about students.
- Furthermore, knowledge of *diagnostic methods* seems to be essential when teachers have to locate possible reasons for errors.

Teachers use these categories of knowledge resources in diagnostic situations as found in the empirical analysis correspond largely with domains of mathematical knowledge which Ball and colleagues propose. The first category concurs with *common content knowledge* and is characterized as a type of general knowledge that is needed in different professions, not only in teaching. However, the second category is knowledge that no other profession needs (*specialized content knowledge*). The third category in addition implies knowledge about students and is conform to the domain *knowledge of content and students*. The last category seems to be rather a type of general pedagogical knowledge (Shulman, 1986, 1987) with being also subject-specific, or even topic-specific. It is a type of mathematical knowledge needed especially in diagnostic situations.

To sum up, the results of the study can be seen in connection with theoretical frameworks that do not only focus on diagnostic competence and therefore can be understood as a specification of such frameworks with respect to the category of diagnostic situations. The empirically found types of mathematical knowledge largely fit into the theoretical framework of Ball et al. (2008) and contain knowledge about methods to localize reasons for errors or misconceptions. A main strategy in diagnostic situations, that is to analyze step by step, resembles the process of decomposing mathematical learning goals Morris et al. (2009) delineate, but goes beyond, for instance, when misconceptions lead to errors that cannot be deduced this way. The results also show that own knowledge plays an important role in diagnostic situations, just as Nickerson illustrates in his model of assessing other peoples' knowledge.

4 Discussion

The main objective of the study was to gain a deeper insight into diagnostic processes of teachers. The broad variety of identified processes shows that informal diagnostic situations make high demands on mathematics teachers. Furthermore, it was possible to point out that different types of (mathematical) knowledge are needed in the above-mentioned diagnostic situations. This may be a reason for different qualities of individual diagnostic processes. This became manifest, for instance, when the participants showed differences in the degree of flexibility, for example, in the number of solution approaches or representations. So, this seems to have an effect on the quality of their analyses in several diagnostic steps. In further studies these differences should be investigated. A possible setting could be to analyze differences between experts and novices with the objective to find indicators for diagnostic competence of mathematics teachers.

Another main objective of the reported study was the identification of different types of mathematical knowledge which teachers use in informal diagnostic situations. The findings provide evidence that in addition to content knowledge, specialized content knowledge and knowledge about content and students, knowledge about diagnostic methods is useful in diagnostic situations.

The present study analyzed diagnostic processes and competences of individuals at a given time in their career. For teacher education, it would be relevant to investigate in what way individuals acquire and develop these diagnostic competences, and which of them can be learned and taught in which way.

Systematic relationships between processes and knowledge resources could not be examined. This may be due to fact that the sample is too small for such analyses. Still diagnostic processes, knowledge resources and their interaction seem to be fundamental for understanding diagnostic competence of mathematics teachers, which involves not only knowledge but also abilities and attitudes. This question is of interest for deriving consequences for teacher education. Further questions arise, for example, how such knowledge can be taught at university.

The model of diagnostic processes is beneficial in that it is possible to classify individuals and so it offers the possibility to compare people or groups of people (e.g., experts and novices). Differences can be made visible, so in further research potential diagnostic types of teachers could be investigated. Furthermore, the identified processes and knowledge resources can be used for the development of concepts in teacher education or teacher further education with fostering of diagnostic competence in mind.

References

Anders, Y., Kunter, M., Brunner, M., Krauss, S., & Baumert, J. (2010). Diagnostische Fähigkeiten von Mathematiklehrkräften und ihre Auswirkungen auf die Leistungen ihrer Schülerinnen und Schüler. *Psychologie in Erziehung und Unterricht*, *57*, 175–193.

Ball, D. L., Lubienski, S. T., & Mewborn, D. S. (2001). Research on teaching mathematics: The unsolved problem of teachers' mathematical knowledge. In V. Richardson (Ed.), *Handbook of research on teaching* (pp. 433–456). New York: Macmillan.

Ball, D. L., Thames, M. H., & Phelbs, G. (2008). Content knowledge for teaching: What makes it special? *Journal of Teacher Education*, *59*, 389–407.

Barth, C., & Henninger, M. (2012). Fostering the diagnostic competence of teachers with multimedia training: A promising approach? In I. Deliyannis (Ed.), *Interactive multimedia* (pp. 49–66). Rijeka, Croatia: InTech.

Brown, C. (1982). A false consensus bias in 1980 presidential preferences. *Journal of Social Psychology*, *118*, 137–138.

Camerer, C. F., Loewenstein, G., & Weber, M. (1989). The curse of knowledge in economic settings: An experimental analysis. *Journal of Political Economy*, *97*, 1232–1254.

Clarke, D. M., Cheeseman, J., Gervasoni, A., Gronn, D., Horne, M., McDonough, A., … Rowley, G. (2002). *Early numeracy research project final report.* Melbourne, Australia: Mathematics Teaching and Learning Centre, Australian Catholic University.

Ericsson, K. A., & Simon, H. A. (1993). *Protocol analysis*. Cambridge, MA: MIT Press.

Fischhoff, B. (1975). Hindsight ≠ foresight: The effect of outcome knowledge on judgment under uncertainty. *Journal of Experimental Psychology: Human Perception and Performance, 1*, 288–299.

Fussell, S. R., & Krauss, R. M. (1989b). Understanding friends and strangers: The effects of audience design on message comprehension. *European Journal of Social Psychology, 19*, 509–526.

Helmke, A., Hosenfeld, I., & Schrader, F.-W. (2004). Vergleichsarbeiten als Instrument zur Verbesserung der Diagnosekompetenz von Lehrkräften. In R. Arnold & C. Griese (Eds.), *Schulleitung und Schulentwicklung. Voraussetzungen, Bedingungen, Erfahrungen*. Baltmannsweiler: Schneider-Verlag Hohengehren.

Hill, H. C., Ball, D. L., & Schilling, S. G. (2008). Unpacking pedagogical content knowledge: Conceptualizing and measuring teachers' topic-specific knowledge of students. *Journal for Research in Mathematics Education, 39*(4), 372–400.

Ingenkamp, K., & Lissmann, U. (2008). *Lehrbuch der pädagogischen Diagnostik*. Weinheim, Basel: Beltz.

Kelley, C. M. (1999). Subjective experience as a basis of "objective" judgments: Effects of past experience on judgments of difficulty. In D. Gopher & A. Koriat (Eds.), *Attention and performance* (Vol. 17, pp. 515–536). Cambridge, MA: MIT Press.

Klug, J., Bruder, S., Kelava, A., Spiel, C., & Schmitz, B. (2013). Diagnostic competence of teachers: A process model that accounts for diagnosing learning behavior tested by means of a case scenario. *Teaching and Teacher Education, 30*, 38–46.

Krauss, S., & Brunner, M. (2011). Schnelles Beurteilen von Schülerantworten: Ein Reaktionszeittest für Mathematiklehrer/innen. *Journal für Mathematik Didaktik, 32*(2), 233–251. Heidelberg: Springer.

Leuders, T., Dörfler, T., Leuders, J., & Philipp, K. (2018). Diagnostic competence of mathematics teachers – unpacking a complex construct. In T. Leuders, K. Philipp, & J. Leuders (Eds.), Diagnostic competence of mathematics teachers. Cham: Springer.

Mayring, P. (1983). *Qualitative Inhaltsanalyse: Grundlagen u. Techniken*. Weinheim, Basel: Beltz.

Morris, A. K., Hiebert, J., & Spitzer, S. M. (2009). Mathematical knowledge for teaching in planning and evaluating instruction: What can preservice teachers learn? *Journal for Research in Mathematics Education, 40*(5), 491–529.

Nickerson, R. S. (1999). How we know-and sometimes misjudge-what others know: Imputing one's own knowledge to others. *Psychological Bulletin, 125*(6), 737–759.

Ostermann, A., Leuders, T., & Philipp, K. (2017). Fachbezogene diagnostische Kompetenzen von Lehrkräften – Von Verfahren der Erfassung zu kognitiven Modellen zur Erklärung. In T. Leuders, M. Nückles, S. Mikelskis-Seifert, & K. Philipp (Eds.), *Pädagogische Professionalität in Mathematik und Naturwissenschaften*. Springer: Wiesbaden.

Philipp, K., & Leuders, T. (2014). Diagnostic competences of mathematics teachers – Processes and resources. In C. Nicol, S. Oesterle, P. Liljedahl, & D. Allan (Eds.), *Proceedings of the Joint Conference of PME 38 and PME-NA 36* (Vol. 4, pp. 425–432). Vancouver, Canada: PME.

Ruf, U., & Gallin, P. (2005). *Dialogisches Lernen in Sprache und Mathematik. Band 1: Austausch unter Ungleichen. Grundzüge einer interaktiven und fächerübergreifenden Didaktik*. Seelze-Velber: Kallmeyer.

Schrader, F.-W. (2011). Lehrer als Diagnostiker. In E. Terhart, H. Bennewitz, & M. Rothland (Eds.), *Handbuch der Forschung zum Lehrerberuf* (pp. 683–698). Waxmann: Münster, Germany.

Schrader, F.-W., & Helmke, A. (1987). Diagnostische Kompetenz von Lehrern: Komponenten und Wirkungen. *Empirische Pädagogik, 1*, 27–52.

Schwarz, B., Wissmach, B., & Kaiser, G. (2008). "Last curves not quite correct": Diagnostic competence of future teachers with regard to modelling and graphical representations. *ZDM – The International Journal on Mathematics Education, 40*(5), 777–790.

Selter, C. (1995). Zur Fiktivität der, Stunde Null' im arithmetischen Anfangsunterricht. *Mathematische Unterrichtspraxis, 16*(2), 11–19.

Shulman, L. (1987). Knowledge and teaching: Foundations of the new reform. *Harvard Educational Review, 57*(1), 1–22.
Shulman, L. S. (1986). Those who understand: Knowledge growth in teaching. *Educational Researcher, 15*(2), 4–14.
Spinath, B. (2005). Akkuratheit der Einschätzung von Schülermerkmalen durch Lehrer und das Konstrukt der diagnostischen Kompetenz. *Zeitschrift für Pädagogische Psychologie, 19*, 85–95.
Swan, M. B. (2001). Dealing with misconceptions in mathematics. In P. Gates (Ed.), *Issues in Mathematics Teaching* (pp. 147–165). London, England: RoutledgeFalmer.
Tversky, A., & Kahnemann, D. (1974). Judgment under uncertainty: Heuristics and biases. *Science, 185*, 1124–1131.
Wartha, S. (2007). *Längsschnittliche Untersuchungen zur Entwicklung des Bruchzahlbegriffs*. Hildesheim, Germany: Franzbecker.
Wiliam, D. (2007). Keeping learning on track. In F. K. J. Lester (Ed.), *Second handbook of research on mathematics teaching and learning* (pp. 1053–1098). Charlotte, NC: Information Age.

Revealing and Promoting Pre-service Teachers' Diagnostic Strategies in Mathematical Interviews with First-Graders

Simone Reinhold

This chapter focuses on pre-service teachers' diagnostic strategies in their analyses of one-on-one mathematical interviews with first-graders. It tackles the question how prospective primary mathematics teachers benefit from university courses which aim at the development of diagnostic competence and how these specific courses should be designed to promote the participants' diagnostic strategies and competence. Taking a domain-specific view on the formative assessment of first-graders' arithmetic prerequisites and learning processes, the Sects. 1 and 2 of the chapter introduces research results on components and types of diagnostic strategies. These research results enrich the discussion about ways of capturing diagnostic competence by an analysis of the details of the diagnostic process. In the Sect. 3 of the chapter, these results and complementary perspectives are taken into account to derive implications for the design of pre-service teacher mathematics methods courses which aim at promoting pre-service teachers' diagnostic strategies in mathematical interviews with first-graders. This discussion also raises questions concerning the qualitative assessment of diagnostic competence in terms of tracking individual pre-service teachers' development of their diagnostic strategies.

1 Diagnostic Competence and Strategies

1.1 Perspectives on the Concept of Diagnostic Competence

Facing and analyzing the diversity, breadth, and depth of young children's mathematical conceptions is an integral element of everyday classroom situations, as primary teachers are challenged to design learning environments which facilitate the

S. Reinhold (✉)
Leipzig University, Leipzig, Germany
e-mail: simone.reinhold@uni-leipzig.de

T. Leuders et al. (eds.), *Diagnostic Competence of Mathematics Teachers*, Mathematics Teacher Education 11, DOI 10.1007/978-3-319-66327-2_7

acquisition of mathematical conceptions. This is referred to as adaptive teaching competence (Wang, 1992) and comprises diagnostic competence since "effective diagnosis is critical for successful teaching" (Helmke & Schrader, 1987, p. 91). Although we find reasonable arguments for a distinction between the terms "diagnosis" and "assessment," these terms are used as synonyms in this chapter: Gathering information concerning students' prerequisites, learning processes or learning results always aims at identifying students' actual learning conditions. With respect to the type of diagnostic situation which is used to gain this information, it is common to distinguish between *initial* assessment (aiming at information about prerequisites or previous knowledge), *formative* assessment (monitoring students' learning with a process-oriented view on the individual), and *summative* assessment (of the learning results). Furthermore, Ginsburg (2009; referring to the work of Jean Piaget) provides a framework for categorizing formative assessments into three groups, namely *observation, test,* and *clinical interview* (see also Sect. 3.1).

As shown in chapter "Diagnostic Competence of Mathematics Teachers Unpacking a Complex Construct" of this volume, studies on the assessment or development of diagnostic competence mainly focus on predictive accuracy of teachers' judgements (cf. Demaray & Elliott, 1998; Südkamp et al., 2012). This predictive accuracy may be measured by numerical indicators concerning the level, the variance or the rank order within classes. Research and discussions in the German-speaking community on mathematics education indicate that diagnostic competence is obviously a domain-specific construct and cannot be generalized (cf. Busch, Barzel, & Leuders, 2015; Prediger, 2010; Scherer & Moser Opitz, 2010; Streit & Royar, 2012; Wollring, 1999). For example, primary teachers who gain appropriate diagnoses on students' conditions in one subject do not necessarily reach the corresponding appropriateness in diagnosing mathematical conceptions (Lorenz & Artelt, 2009).

As stated by Philipp and Leuders (2014, see also chapter "Diagnostic Competence of Mathematics Teachers: Unpacking a Complex Construct") or Reinhold (2014, 2015), there is a lack of knowledge concerning the cognitive processes, resources and diagnostic practices which enable the diagnosing teacher to judge or evaluate an individual student's learning development. Diagnostic practices or facets of diagnostic strategies in initial or formative assessment situations have only scarcely been studied, so far. Furthermore, merely focusing on the quantitative measurement of diagnostic accuracy places specific constraints on capturing pre-service teachers' diagnostic competence: Judgments on the *level* of a task might closely rely on theoretically acquired knowledge on learning trajectories or students' typical (mis)conceptions. In contrast, the *variance* or the *rank* (e.g., relative strengths and weaknesses of a class) deeply depends on previous practical experience with the students in question. Novices like pre-service teachers usually do not have this knowledge at hand when they meet their class of students for the first time.

To cope with the demand of developing adaptive teaching competence, (prospective) teachers need to have a wide variety of knowledge, strategies and practices at their disposal. Based on the domains defined by Shulman (1986), Ball, Sleep,

Boerst, and Bass (2009) suggest to integrate the capability of "eliciting and interpreting individual students' thinking" into the set of "high-leverage practices." Novices should be familiarized with these "high-leverage practices" which operationalize teachers' *pedagogical content knowledge (PCK)*. Moreover, these practices refer to *knowledge of content and students (KCS)* as subdomain of *PCK* (Ball, Thames, & Phelps, 2008, p. 403). In this sense, analyzing an individual student's mathematical concepts is likely to contribute to a deeper understanding of common (mis-)conceptions. Teachers' analyses of individual cases are expected to help to develop *KCS*, they may improve a teacher's own practices in terms of attention or "noticing" (Jacobs et al., 2010), and thereby enrich his or her diagnostic expertise. Yet, research concerning the quality of pre-service teachers' "eliciting and interpreting," on their (intuitive) diagnostic practices or capability of referring to *KCS* is rather rare. A promising starting point for uncovering these qualitative facets of diagnostic strategies is a closer look on (prospective) teachers' ability "to notice": Teachers' attention and their capability to focus attention tend to be crucial individual prerequisites for well-considered acting within a diagnostic situation. Attending as an integral element of "professional noticing of children's mathematical thinking" defined by Jacobs et al. (2010, p. 172) refers to the skill of "being able to recall the details of children's strategies". In line with these considerations, Sleep and Boerst (2012, p. 1039) conceptualize this particular "high-level practice" as a subcomponent of the domain "assessing student thinking."

Making an effort to learn more about teachers' professional noticing, Barth and Henninger (2012) focused on teaching situations which were shared via videotaped classroom sessions (see also Sect. 3.2). After watching the sessions, the participants (pre-service teachers) completed an online questionnaire related to their sources of information. Results from the analyses of this data show that the pre-service teachers rely on at least four distinct sources: *observable situation-dependent information* (e.g., body language, reaction on classmates), *non-observable situation-dependent information* (e.g., interpretation and stereotypes about personal traits), *class-specific information* (in one single case) and *professional or experimental knowledge* (e.g., knowledge which helps to judge if a student's answer is adequate). The study conducted by Barth and Henninger (2012) also shows:

> that it is not easy for pre-service teachers to describe the observations that lead to their diagnoses. Often, they described their interpretations instead, which were not linked to a specific perception (p. 60)

Schack et al. (2013, p. 387; see also Sect. 3.2) identified themes that emerge from the analyses of pre-service teachers' noticing in their analyses of excerpts from diagnostic interviews, as well. They distinguish between *identifying key activities*, *identifying additional activities*, *presumptions* concerning the child's computation, *purporting evidence,* and *cognitive interpretations*. In addition, they exemplify that professional noticing, in terms of attending accurately, is likely to facilitate the achievement of sound interpretations of a child's mathematical concepts.

1.2 Process-Oriented Approaches to Diagnostic Competence

The set of interrelated skills labeled as "professional noticing" by Jacobs et al. (2010) also includes "interpreting children's understanding" and being able to make decisions on "how to respond on the basis of children's understanding." This hints at various elements of "professional noticing," and, additionally, reveals that noticing, interpreting and interacting are parts of a multidimensional cognitive process. Especially in the domain of primary mathematics education research, research studies *and* everyday formative assessment in the classroom intensely deal with many aspects of (young) children's heterogeneous abilities and developments. Diagnostic expertise also comprises aspects that are somewhat vague and difficult to capture, like diagnostic sensitivity, curiosity, an interest in children's learning processes or the aptitude to gather and interpret relevant data in non-standardized settings (Prediger, 2010). Hence, research on diagnostic expertise of primary (pre-)service teachers has to reach beyond measuring accuracy in predicting children's scores in assessments.

Klug (2011) and Klug, Bruder, Kelava, Spiel, and Schmitz (2013) offer a general model which is in line with this process-oriented attitude towards diagnostic competence: They define a *pre-actional phase* (including considerations of preparing diagnostic activities or the choice of diagnostic tools) which is followed by an *actional phase* (where data collection and interpretation takes place). The latter is followed by a final *post-actional phase* which implies taking the necessary action from the data collection and interpretation. This may help to design a specific support for a student or evaluate a previous program for fostering. In this sense, diagnosing and fostering may be part of a repeated diagnostic process and labels the *diagnostic macro process* for the study referred to in this chapter.

Following this general model, it can be assumed that each of these phases is distinctively characterized by a set of *diagnostic micro-processes*. The focus in this research is on investigating the elements for determining the *actional phase*. In this setting, preparing a one-on-one diagnostic interview usually takes place in the *pre-actional phase* as the teacher decides which questions or tasks should be used in the interview. The cognitive processes guiding the teachers' choice of tasks may play an important role. In the *actional phase*, conducting the interview is in the focus: Here, micro-processes of initially interpreting the gathered information may take place. Additionally, drawing first conclusions while conducting the interview or capturing the (in)correctness of students' responses may take place in this phase. Obviously, collecting data, interpreting, and drawing further conclusions are likely to be based on different kinds of knowledge (e.g., *KCS*, see Fig. 1) which may have deep impact on a (pre-service) teacher' diagnosis derived from the interview. For example, students' conceptions "must be reconstructed by interpreting their utterances" (Prediger, 2010, p. 76) as there is no direct access for the interviewer. Yet, we have little knowledge on characteristics of this interpretation or on details of "gathering information" (Klug et al., 2013, p. 39), so far.

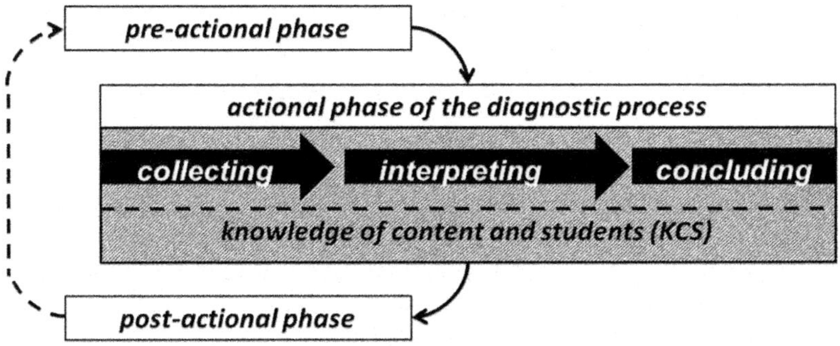

Fig. 1 Differentiating the micro-processes in the actional phase of diagnosing (Reinhold, 2014, p. 43; for the general model see also Klug, 2011, p. 17)

In this sense, domain-specific diagnostic competence needed for the initial and formative assessment of first-graders' mathematical conceptions and learning processes includes the willingness and the cognitive capacity to actively analyze children's mathematical prerequisites and learning processes. This includes meta-knowledge about facets of diagnostic strategies, and the awareness and availability of diagnostic strategies in situations of initial or formative assessment. Yet, applying an appropriate diagnostic strategy in the actional phase of diagnosing can be regarded as a "high-leverage diagnostic practice" which needs to be specifically promoted in teacher education. In this sense, a deeper analysis of varieties of diagnostic strategies and the micro-processes determining the actional phase of diagnosing during diagnostic interviews appears to be a promising approach.

As pointed out above, (prospective) primary mathematics teachers should be sensitized for variations, range, and depth of young children's mathematical thinking and capable of conducting initial and formative assessment. In this sense, preparing, conducting, and analyzing students' mathematical conceptions in one-on-one interviews offers substantial learning opportunities and supports the development of prospective teachers' (PTs') diagnostic attitude (Peter-Koop & Wollring, 2001; Prediger, 2010; Sleep & Boerst, 2012, see also chapter "Diagnostic Competence for Dealing with Students' Errors – Fostering Diagnostic Competence in Error Situations" for more details). Yet, qualitative facets of the diagnostic strategies during such one-on-one diagnostic mathematics interviews have only been scarcely studied so far: What characterizes (prospective) mathematics teachers' diagnostic strategies when they diagnose young children's individual mathematical approaches in one-on-one interviews? Which types of diagnostic strategies can be reconstructed from analyses of (prospective) mathematics teachers' analyses of video-taped diagnostic interviews? What kind of knowledge (e.g., *KCS*) is informing their diagnosis?

2 Analyzing Diagnostic Strategies

2.1 Methods: Uncovering Individual Facets of Diagnosing

Aiming at a more process-oriented approach to diagnostic competence and in search of an empirically grounded framework for a qualitative view on PT's cognitive activities during formative assessment, the main purpose of previous studies within the author's project *diagnose:pro* has been to detect details of diagnostic strategies. With a focus on PTs' diagnoses of individual arithmetic approaches of first-graders, the author and her team tried to capture the variety of diagnostic strategies. Here, we report on various challenges we faced concerning the choice of appropriate methods, and describe the variation of methods we finally arrived at.

Norton, McCloskey, and Hudson (2011) suggested using video-vignettes to gain a more process-oriented insight into prospective teachers' assessment of students' mathematical thinking. In a so-called prediction assessment (Norton et al., 2011, PTs analyzed a video of one child solving a mathematical task. In the next step, they were asked to predict the child's response to the following task and give a written record of their prediction. According to their findings and experiences with this method, Norton et al. (2011) considered video-vignettes to be a substantial tool either for teacher instruction (see also Sects. 3.1 and 3.2) as well as for the purpose of assessing:

> Video clips provide an ideal medium for assessing teacher knowledge because they are both replicable and authentic, in that they depict real students and teachers engaged in doing mathematics (Norton et al., 2011, p. 308).

Building a model of a student's mathematical conception in these settings is expected to initiate evidence-based explanations for students' reasoning (p. 320) and assumed to be useful for reaching appropriate predictions (p. 321). Yet, the authors also mentioned that a refinement of this assessment should make the PTs face novel situations in a post-assessment of their growth when predicting students' responses.

Using an adapted version of the "prediction-assessment" between 2011 and 2012, we asked PTs of several mathematics methods courses in their last year of university studies (Master of Education) to give written records of their predictions after watching a video-clip. These assessments (with excerpts from a diagnostic video conducted by the author with one first-grader) were incorporated in the first and the last sessions of the course. During the semester, the pre-service teachers prepared, conducted, and analyzed individual diagnostic interviews with several first-graders. Analyses of those written documents from the beginning and the end of the course were most interesting as they indicated PTs' development, for example, demonstrated in richer details of their prediction. Obviously, they made use of the *KCS* and issues they had been studying throughout the semester. For example, they took into account details of theoretical models concerning the development of counting strategies or evolving strategies of addition and subtraction in the range from one to 20. Yet, we also encountered constraints concerning our interest in the processes of diagnosing or in the analysis of a one-on-one interview, respectively: Analyzing a collection of *products*, we missed details on the *processes* the PTs had gone through while achieving their predictions.

Based on these experiences and in search of insights into ongoing cognitive processes during the analysis of a one-on-one interview, we shifted our design for subsequent studies: During 2013, seven pre-service teachers agreed to take part in retrospective interviews. They were asked to comment on the video-recording of an interview they had conducted shortly before and stop the video at any time in order to "analyze the interview." When the PTs stopped the video, we expected them to mention any observation they considered to be remarkable and state a diagnosis they might derive from this specific situation they observed in the interview. The interviewer encouraged PTs to explain what kind of knowledge, information or evidence warranted the PT's uttered diagnosis if comments were rather short or pure in detail. The retrospective analyses of diagnostic interviews by the PTs offered the chance to get a process-oriented insight into diagnostic strategies in a "biplane" diagnostic situation—with a diagnosis of a child's mathematical conception (by the PT) and a diagnosis of the PT's diagnostic strategies (by the interviewer) that took place at the same time (for more details, see Reinhold, 2014, 2015). This method resembles a design chosen by Wilson, Mojita, and Confrey (2013) who paused the watching of a video-recording of an interview (with the student Emma) periodically so the PTs:

> could respond to questions aimed at eliciting their hypotheses of Emma's understanding and predicting the ways that she would use that understanding on a subsequent task (p. 110).

The analyses of the data and methodology were based on grounded theory. This comprised open, axial and selective coding and constant comparison (cf. Corbin & Strauss, 2008) which was supported by the software Atlas.ti. In a first step, PTs' verbal utterances and further observations (e.g., frequency of pausing, connections between several stops in one scene) were coded in an open process. These codes supported the identification of sub-categories of collecting, interpreting, and concluding within the actional phase of a diagnostic interview and thereby contributed to the refinement of elements of PTs' diagnostic strategies.

2.2 *Results: Elements of Pre-service Teachers' Diagnostic Strategies*

The more we can learn about details of PTs' diagnostic strategies, and the more we are able to make details of the underlying cognitive processes explicit, the better we can foster PT's diagnostic strategies and practices in the concrete situation of formative assessment:

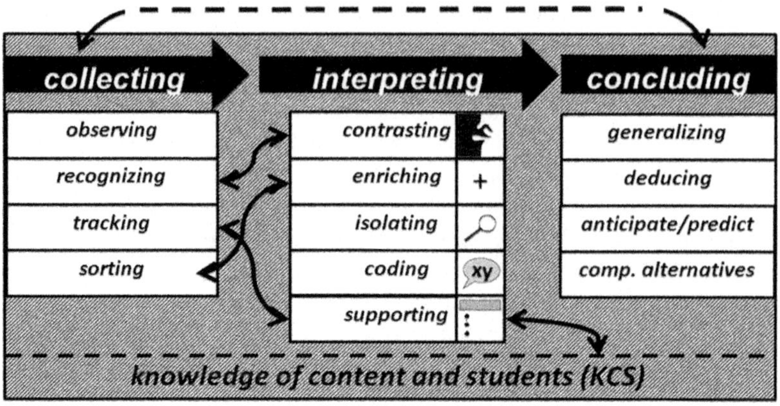

Fig. 2 Sub-categories of collecting, interpreting, and concluding

> To foster diagnostic competence, it is crucial to know which information or knowledge sources play the most important role during the process of diagnosing students' learning prerequisites (Barth & Henninger, 2012, p. 50).

With a strong focus on a process-oriented perspective on diagnostic competence, it is also useful to find out more about the way in which any information and the knowledge sources are used during a diagnostic process. Our findings underpin that cognitive elements of PTs' diagnostic strategies during one-on-one interviews often resemble processes in qualitative data analysis. This includes acts like *collecting*, *interpreting*, and *concluding* within the retrospective process of analyzing a child's mathematical conception from a one-on-one interview (see Fig. 1). Beyond this, more detailed findings identify sub-categories of *collecting*, *interpreting* or *concluding* (displayed in Fig. 2). Those sub-categories and the interrelations between them reveal and help to describe distinct types of diagnostic strategies.

2.2.1 Facets of Collecting Data

We investigated qualitative details of pre-service teachers' data collection and found that micro-processes of collecting within the actional phase are characterized by various types of data collection and concerning the choice of information (see Reinhold, 2015).

For example, we coded PTs' data collection as *observing* when they watched closely what was happening in the diagnostic situation and listened attentively to the child's utterances. In general, the PTs paid attention to significant details, but they also rather frequently noticed the (singular) occurrence of micro-incidents which were only loosely connected (more or less collecting "a colorful bunch of flowers"). Collecting on a higher level was found in the sense of *tracking*

Table 1 Various sources for interpretation: What is collected?

Collected	Example
Verbal utterance	"This boy, he was able to identify the summands and he said 'This number and this number equals this number." (Anne)
Activity	"He's drawing a circle around this piece of the pattern." (Pam)
(In)correctness of solution	"He was supposed to draw a circle around repeating parts of the pattern, but he failed." (Pam)
(Elements of) strategy	"He used counting strategies, seeing four and continued counting from that first summand." (Sue)
Eye movement	"He hesitated and looked the other way." (Anne)
(Subtle) movements of lips, head, hands	"I see he is nodding and I guess he's counting up to five here." (Lisa)
Emotional state	"I got the impression he'd start crying." (Anne)
interviewer's behavior	"Okay, I liked what I did in this situation as we decided to accept 'wrong' answers, too." (Sue)

as pre-service teachers followed a series of activities or utterances over a longer sequence. In one of the re-interviews, for example, the master student, Lisa, comments on an interview with six-year old Sam (Reinhold, 2015). He is asked to take five counters (one side blue, the other side red) and reason about possible ways of displaying various equations for reaching the sum of five. Sam starts with spreading the counters on the table before he begins sorting them ("Three red ones and two blue ones"). Here, Lisa stops the video:

Lisa (01:51): "To comment on this, I'd say he separated red and blue from the beginning and named what was lying on the table."

Later on, Lisa tracks this idea and collects further information from subsequent situations which refer to this issue (sorting and considering position of colors).

Lisa (02:16): "Here, it is clear that he separated the colors from the beginning."

Lisa (10:20): "We wanted them to find that *sorting* the possible additions helps to find all of them, yes and he is arranging them in *any* kind of structure, but... not the one we had intended them to find, (...). But in a way he *does* sort the possible arrangements because in this corner here, the blue ones are closer together. In the next row, the blue ones stick closely together, too, and there the red ones."

When single incidents or details were repeatedly identified, we coded this as *recognizing*. *Sorting* in data collection was identified when PTs found (or intentionally searched for) groups or patterns in children's utterances or actions.

A further analysis of pre-service teachers' comments also reveals that the range of mentioned details is wide (verbal utterances, activities, (in)correctness of solution, [elements of] children's strategies or eye movements, see also Reinhold, 2015 and Table 1).

2.2.2 Interpreting and Concluding

PTs micro-processes of interpreting and concluding in the actional phase of diagnosing are characterized by a variety of subordinate elements, as well. We frequently found comments among the PTs' utterances where they *compare* details to a child's previous utterances or actions ("a moment ago"). It is also fairly common for PTs to compare their observations to utterances or actions they have observed in previous diagnostic interviews with other children ("just like Lisa in the interview before") or their own concept.

Comparing sometimes even occurred in terms of direct *contrasting* different scenarios. For example, Ann, a master student in her last year of studies, comments on her interview with the first-grader, Ben, who is asked to count chestnuts and put them into boxes (for more details see also Reinhold, 2014). In the re-interview Ann states:

> Here, he saw, okay, there are four in one box and there are another four in the second box, well, four plus four equals eight, but he didn't do it that way in the next task. There he'd count single ones, it was done quite differently.

Furthermore, novices try to grasp unfamiliar, but obviously central aspects of a child's conception by *coding* the phenomena they observe. They often refer to these "codes" later on in the re-interview and thereby substitute established terms (e.g., "shortcut" instead of "subitizing" in counting tasks).

2.2.3 Types of Strategies

Pre-service teachers' individual diagnostic strategies refer to different elements of the exemplified sub-categories of collecting, interpreting, and concluding which are displayed in Fig. 2.

In this sense, individual diagnostic strategies may be captured by referring to these sub-categories and describing how elements of collecting, interpreting and concluding in the PTs' diagnostic process are intertwined (indicated by the arrows which connect the sub-categories). Taking a closer look on common types of strategies, the strategy descriptive collector can be observed when the PTs focus on collecting and describing the child's actions and neglect both interpreting and concluding. A concluding collector strategy is characterized by skipping elements of interpretation. In this strategy, collecting directly leads to conclusions.

As the arrows in Fig. 2 indicate, these diagnostic strategies are far from operating like linear processes, leading directly from collection via interpretation to conclusion. Instead, pre-service teachers may run through these micro-processes in circles: For example, they may find out during the diagnostic process that they have not collected enough or sufficient data for contrasting (sub-category of interpreting) and therefore intentionally return to collecting—a type of diagnostic strategy we call a branched interpretation.

2.2.4 Discussion: Facets of Diagnostic Strategies and Types

The results of our studies in the project *diagnose:pro* underpin the notion that cognitive elements of PTs' diagnostic strategies often resemble processes we encounter in qualitative data analysis, like Jungwirth, Steinbring, Voigt, and Wollring (2001) or Prediger (2010) have pointed out in previous studies. For example, our findings concerning facets of data collection (namely *observing*, *tracking*, *recognizing*, and *sorting*) are consistent with results presented by Barth and Henninger (2012). Concerning sub-categories of interpreting and concluding, Wilson et al. (2013) observed similar incidents as they report:

> Teachers use the processes of Describing, Comparing and Inferring to construct models of student thinking. As these models are incorporated into their own knowledge of content and students, teachers are reconstructing their knowledge (p. 108).

The review of literature also points at parallels to the diagnostic strategies we revealed. For example, our findings of the strategy *concluding collector*, resemble an observation stated by Bräuning and Steinbring (2011, p. 931) who describe how a teacher "springs spontaneously to an immediate conclusion" in her diagnostic interaction with a child. These observations reveal diagnostic practices which have to be uncovered in order to make them explicit. Hence, deepening the knowledge about diagnostic strategies can at the same time contribute to promote PTs diagnostic expertise. Therefore, Sect. 3 takes a closer look at results on features of courses which are assumed to foster diagnostic competence in a more general way (Sects. 3.1 and 3.2). Concrete implications we drew from these discussions to modify our mathematics methods courses include the idea of making diagnostic strategies explicit in the sense of teaching elements of diagnostic strategies during the course (Sect. 3.3).

3 Promoting Diagnostic Strategies and Expertise in Pre-service Teacher Training

3.1 Features of Supportive Courses for Practicing Teachers

High-quality professional development engages teachers in concrete tasks (e.g., tasks of assessment or observation) and focuses on students' learning processes (Borko, Jacobs, & Koellner, 2010). Being engaged in diagnostic processes most often starts with collegial communication among teachers about their students' content knowledge or learning processes. Sources or reasons for this diagnostic collegial communication and interpretation are often written products (e.g., tests, drawings, and posters), verbal articulations (e.g., comments during group-discussions), or things teachers observe during their students' work as they carefully monitor students' learning processes (Clarke & Wilson, 1994). Bräuning and Steinbring (2011) enhanced the positive effects of collegial reflection on the development of diagnostic competence as they designed a course for practicing teachers which comprised several mathematics diagnostic talks with individual children. They characterize these diagnostic talks in the following way:

> In mathematical diagnostic talks, the teacher tries to investigate the particularities of a child's mathematical knowledge, imagination and ways of proceeding. The setting of a diagnostic talk as a one-on-one situation offers possibilities for the teacher to intensively turn towards one student and to 'scout' out his/her understanding of mathematical problems (Bräuning & Steinbring, 2011, p. 928).

This description and intention is in line with what is most often called a "clinical interview"—a method introduced by Jean Piaget who elaborated this approach to the so-called revised clinical method (Ginsburg & Opper, 1998). Since then, this method comprises questioning and observing a child in a one-on-one situation in combination with using hands-on materials to investigate children's cognitive development in various (mathematical) domains (cf. Ginsburg, 2009). Thus, we use the terms "clinical interview", "diagnostic interview", "one-on-one interview," or "diagnostic talk" as synonyms in this chapter. A particular significance of the clinical method for practicing teachers derives from the opportunity for teachers to:

> apply adapted problems and tasks devised originally for research purposes, and with assistance, begin to make connections with theory and their own practice (Hunting, 1997, p. 146).

Thereby, the clinical interview is said to "open the door for teachers to begin to expand their experience of how children's minds work mathematically" (Hunting, 1997, p. 161).

Not surprisingly and in consideration of this background, it is widely acknowledged that preparing, conducting, and analyzing one-on-one interviews promote teachers' diagnostic awareness. For example, research-based frameworks (e.g., concerning learning trajectories on the early development of basic arithmetic competencies) resulted in the design of standardized task-based interviews which help teachers to assess children's thinking in the context of mathematics learning in school. This has been recorded, for example, for the *Early Numeracy Research Project*'s task-based assessment interview (ENRP; e.g., Clarke, Clarke, & Roche, 2011; Clarke, 2013), within the project *Count Me In Too* (CMIT; e.g., Bobis et al., 2005) or whilst using the German adaption of the ENRP in the *Elementar Mathematisches Basis Interview* (EMBI; Peter-Koop, Wollring, Spindeler, & Grüßing, 2007). Obviously, we can state a certain "power of one-on-one interviews" (Clarke, 2013; chapter "Supporting Mathematics Teachers' Diagnostic Competence Through the Use of One-to-one, Task-Based Assessment Interviews" in this volume).

The use of video-clips to get teachers engaged in diagnostic situations has become a fairly wide-spread method in teacher education, as an alternative to conducting clinical interviews with each student. For example, Whitenack, Knipping, Novinger, and Stanfifer (2000) adapted a design introduced by Lampert and Ball (1998) which provided primary teachers with videotaped excerpts from arithmetic interviews. Here, teachers were asked to develop so-called mini-cases which offered the opportunity to engage in "second-hand" inquiry, and, in the end of the course, enabled them to give sound hypotheses on children's thinking.

3.2 *Promoting Pre-service Teachers' Diagnostic Expertise*

Scholars have recommended addressing the variety of students' individual (mis-) conceptions in pre-service teacher education for a long time (e.g., Tirosh, 2000; McDonough, Clarke, & Clarke, 2002). Actively engaging pre-service teachers in observations of children at work (e.g., clinical interviews or analyzing video-recordings) is a recommended approach.

Hence, preparing, conducting, and analyzing one-on-one interviews also provide novices with substantial learning opportunities as they study students' mathematical conceptions (cf. Prediger, 2010; Sleep & Boerst, 2012). Developing a sensitive diagnostic attitude is also supported by involving pre-service teachers in research projects that include interview assessments (cf. Jungwirth et al., 2001; Peter-Koop & Wollring, 2001). For example, Jungwirth et al. (2001) stress the benefit pre-service teachers derive from being active members of a research team undertaking interpretative classroom research, and being involved in data collection where they design, document, and analyze diagnostic interviews. This is said to enhance their individual mathematics education background since they may recognize "the universal in the special case" (Jungwirth et al., 2001, p. 53).

Substantial diagnoses derived from clinical interviews or diagnostic talks initially depend on the choice of questions (Hunting, 1997, p. 153). The task-based interview assessments discussed in Sects. 1 and 2 (e.g., ENRP) take this responsibility for choosing questions and analyzing students' answers: The design and analyses of the interview data is based on research results concerning learning trajectories. The interviewer simply used the provided script directory for both interview questions and analyses. So, on one hand, working with prepared interview tools may serve as a sound method for achieving diagnostic sensitivity. On the other hand, time and resources to conduct prepared task-based interviews are not always available in everyday classroom situations where informal formative assessment may be needed.

Scholars in various settings have analyzed the pre-requisites pre-service teachers mostly bring with them to cope with the specific challenges of formative assessment, and also hint at constraints we have to face when educating the prospective educators. For example, Moyer and Milewicz (2002) report on pre-service teachers' difficulties in asking the right questions as they conduct an interview with a child. They argue that the use of appropriate questioning strategies is essential for assessing mathematics. Most of the pre-service teachers in their sample prepared several interviews (varying in length) with children aged from five to 12 and submitted a selected interview for subsequent reflection. In preparation of the reflection, they transcribed the entire audio-taped interview and recorded noteworthy details.

Their reflections consisted of written comments which were achieved through guided analysis. Data analyses led to three general categories of questions used by the pre-service teachers, namely check listing, instructing (rather that assessing) and probing (with follow-up questions). Instructing instead of assessing tended to be one of the wide-spread strategies followed by the pre-service teachers. Furthermore, Moyer and Milewicz report on several pre-service teachers who merely attempted to teach mathematical conceptions instead of trying to elicit children's individual concepts. To meet and overcome these difficulties, they stress the importance of intense reflection on the used questioning strategies (see Sect. 3.3.1).

Barth and Henninger (2012) concluded that previously gained knowledge concerning the class or single students tends to be less influential on the pre-service teachers' diagnoses than the actual situation. Hence, they suggest:

> (…) a learning environment which focuses mainly on situative cues could be a useful way to foster the ability to make a competent diagnosis in teaching situations (p. 59).

Similar to Whitenack et al. (2000) or Lampert and Ball (1998) with in-service teachers (see Sect. 3.1), Schack et al. (2013) use video-excerpts from clinical interviews with children as "representations of practice." However, the pre-service teachers have an opportunity to conduct similar diagnostic interviews themselves (labeled as "approximation of practice" by Schack et al.):

> Discussions around the video-clips required PSETs (pre-service elementary teachers, note S. R.) to not only attend and interpret, but to make decisions about next diagnostic or instructional steps (Schack et al., 2013, p. 385).

3.3 Implications for the Design of Courses for Promoting Pre-service Teachers' Diagnostic Expertise

In relation to the development of *PCK* and diagnostic competence, numerous researchers suggest to integrate opportunities to analyze students' errors or develop a task, which students may use to express their conception in an informal way (e.g., Kilic, 2011; Wollring, 1999). An implementing phase of preparing, conducting and analyzing one-on-one interviews helps pre-service teachers to get an idea of diagnostic situations they will have to cope with in class, later on (cf. Peter-Koop & Wollring, 2001; Peter-Koop, 2006). Having shared these experiences from corresponding activities in pre-service teacher education for some years, the author and her research team tried to elaborate the design of our courses through time. Hence, Sects. 3.3.1, 3.3.2 and 3.3.3 will highlight implications and conclusions we drew from experience and literature review, and give an idea of and arguments for a program designed for two semesters, namely the teaching project *MathWerk*[1].

---

[1]The project *MathWerk* received funding for the implementation of an innovative teaching project in university (BMBF, LaborUniversität Leipzig, project "StiL – Studieren in Leipzig", from October 2015 to September 2016).

The course in the project is designed as a spiral with diverse opportunities to develop pre-service teachers' individual diagnostic strategies. All activities of the course are connected to any of the following *three main ideas*:

- Action and reflection of experienced action
- Support PCK with a focus on KCS
- Explicit analyses of personal preconditions

3.3.1 Action and Reflection on Experienced Action

Referring to the cyclic character of the process-model of diagnosing introduced by Klug (2011) and Klug et al. (2013), acting and reflecting on experienced action tend to be key factors for the development of diagnostic competence. Hence, acting and reflecting can be assumed to shift or even change diagnostic strategies, as well.

In this sense, Moyer and Milewicz (2002) argue that pre-service teachers benefit most from conducting and analyzing clinical interviews "by scrutinizing their own performance and reflecting on the questions they use in these interviews" (p. 294). Moreover, they suggest guiding pre-service teachers through their self-reflection process in their pre-service teacher education courses as this is expected to facilitate pre-service teachers' recognition of effective and ineffective questioning strategies. In their own studies, Moyer and Milewicz (2002, p. 298) additionally provided video-data on the course instructor's interviews with children. Specific questioning strategies used by the instructor in those interviews were highlighted in the video's analysis. Barth and Henninger (2012, p. 60) emphasized this kind of reflection and suggested the creation of suitable multimedia-based learning scenarios to support pre-service teachers.

Based on our previous research and on these considerations, we integrated the following *activities* in various settings throughout the project:

- Reflection on the video of a mathematics diagnostic interview (conducted by another pre-service teacher) with discussions on the child's conception and the pre-service teacher's strategy throughout this interview.
- Personal exploration and design of a learning environment which is expected to be suitable for formative assessment (see also Sect. 3.3.2).
- Planning, conducting and reflecting diagnostic interviews with several children (pre-service teachers in groups of three or four), embedded in a learning environment.

3.3.2 Supporting PCK with a Focus on KCS

Concerning the reflection on experienced action in the field of diagnosing, Klug (2011, p. 21) indicates the importance of knowledge. This can be seen not only in the light of diagnostic methods and facets of *pedagogical content knowledge (PCK)* in general, but also concerning *knowledge of content and students (KCS)* and *specialized content knowledge (SCK)* for specific mathematical issues. In line with this demand, courses provided by Fischer and Sjuts (2012), for example, include both theoretical background and practical exercises for developing diagnostic talks. Here, students' written responses are analyzed during the course, which provides the starting point for ensuing considerations for fostering this specific student. To specify the theoretical background which is expected to be useful for developing diagnostic competence, we refer to learning trajectories (LTs). These research-conjectured and empirically based articulations provide substantial knowledge on the ways in which students' progress from informal to more sophisticated conceptions (see also Sect. 3.1). Wilson et al. (2013), for example, successfully utilized LTs on rational number in settings addressing practicing and prospective elementary teachers, and concluded that knowledge of a LT supports pre-service or in-service teachers in reconstructing students' conceptions and reasoning (cf. capturing LTs in "growth points" developed for the interviews assessments in the *ENRP*, *CMIT* or *EMBI*; see Sect. 3.1).

To promote and further evaluate the domain-specific features of diagnostic competence in a setting focusing on first-graders, the emphasis of the course we report on here is on young children's arithmetic competencies. As this focus still comprises a wide range of aspects in the field of early arithmetic, we selected the following three topics which all relate to basic ideas of great importance during the first weeks and months in primary mathematics education:

- Counting strategies and enumeration via (quasi-)subitizing
- Seriation and relation of numbers (1–20)
- Part-whole relationships (numbers 1–20)

To ensure that the pre-service teachers had the basic mathematical knowledge for diagnosing mathematical basics, these topics (e.g., cardinal and ordinal aspects of numbers, basics of set theory) were revisited during the first phase of the semester. Rather than doing this revision in the traditional manner using lectures, the pre-service teachers were introduced to the concept of substantial learning environments (cf. Hirt & Wälti, 2008). On one hand, exploring these learning environments (e.g., so-called arithmogons suggested by McIntosh & Quadling, 1975; Wittmann, 1981) enabled the pre-service teachers to discover the specific potential provided by the "low level – high ceiling" characteristics of selected learning environments. As they tried to find and work out more elaborated variations themselves (using higher numbers, fractions instead of natural numbers, etc.), they had to remind themselves of connected mathematical basics and principles underlying the structure of a learning environment (e.g., the underlying set of equations in an arithmogon). They also reflected on their own learning experiences in group discussions and drew didactic conclusions for using these learning environments in the classroom. On the other hand, doing so provided an ensuing diagnostic activity where they experienced learning environments and adapted them for preparing, conducting and analyzing one-on-one interviews with first-graders.

Before preparing the one-on-one interviews (see also Sect. 3.1), the pre-service teachers are introduced to research results concerning the chosen topics (e.g., common

[mis]conceptions or students' concept development for addition strategies) in order to enrich their *KCS*. They receive an introduction into the method of task-based mathematics interviews (e.g., EMBI; Peter-Koop et al., 2007) and intensely discuss the question how the learning environments could be embedded in a clinical interview. The preparation of their own interviewing activity also includes a discussion of concrete examples for elements of the subcategories of collecting, interpreting and concluding. This is connected to the presentation of little icons (see Fig. 2; "diagnostic apps" to illustrate sub-categories of interpreting). Explicit labeling, providing knowledge on these elements and their sub-categories and helping to memorize crucial elements of a diagnostic micro-process is expected to provoke awareness in ensuing diagnostic situations. These "diagnostic apps" might be able to serve as for eliciting and interpreting students' thinking (Sleep & Boerst, 2012).

3.3.3 Explicit Analyses of Personal Preconditions

As previously mentioned, further variables concerning personal preconditions or resources are evident: What kind of professional self-concept is developed by a pre-service teacher? Is he or she motivated to diagnose and curious to find out more about the child's mathematical conceptions? Is he or she confident in terms of self-efficacy in diagnosing? What attitude towards diagnosing appears?

Prediger (2010, p. 76) stressed the importance of an interest in students' thinking, which should not be mixed up with a deficit-oriented curiosity about "simple studies of errors and misconceptions." Instead, this is an interest in individual learning processes, ideally associated with an interpretative attitude and the willingness to undergo deep interpretative analyses (cf. Jungwirth et al., 2001; Peter-Koop & Wollring, 2001). Therefore, we developed a questionnaire trying to capture and document the participating pre-service teachers' prerequisites and their personal attitude towards children's (mis-)conceptions.

4 Conclusions for Prospective Research

Designing specific courses to enhance pre-service teachers' diagnostic sensitivity and their competencies in eliciting and appropriately analyzing children's mathematical conceptions raises further questions concerning the evaluation of such courses. Is it possible to state and analyze in detail the development of diagnostic strategies? How could we measure this development concerning rather vague aspects in a qualitative way, facing the difficulties discussed above (see Sect. 2.1)?

For prospective research, we suggest a qualitative investigation which intends to track individual pre-service teachers' development of their diagnostic strategies. In our present studies, pre-service teachers in two differently designed university courses are asked to take part in the study. The courses are taught by the same lecturer and share a common core of contents as they both tackle arithmetic and mathematics methods for early primary school. However, the pre-service teachers of the test group engage in an intervention which consists of discussions of weaknesses and strengths in other novices' analyses of one-on-one interviews, or an intense work on preparing, conducting and analyzing their own interviews with young children. This pre- and post-design of the ongoing studies resembles the research by Schack et al. (2013) who used their framework for assessing pre-service teachers' professional noticing in a pre- and post-assessment as they exemplify individual qualitative shifts in pre-service teachers' noticing (see Sect. 2.2.1). Wilson et al. (2013) used a similar approach as they tried to identify practicing and prospective individual teachers' transitions in their processes by describing and interpreting the observations of an interview situation.

Based on these considerations and our initial data collection in the project *MathWerk*, we intend to capture the diagnostic strategies used by the participating PTs during the analyses of a clinical interview in the beginning of the course. Here, the empirically grounded framework described earlier (see Sect. 2.2) helps to identify elements of the strategies and grasp strategies of individuals. Comparing the characteristics of these strategies to strategies the PTs use later in the project appears to be a meaningful way to document qualitative facets of the development of their diagnostic competence. This detailed documentation may also serve the instructor's evaluation of the course in the sense of an assessment.

References

Ball, D. L., Sleep, L., Boerst, T., & Bass, H. (2009). Combining the development of practice and the practice of development in teacher education. *The Elementary School Journal, 109*(5), 458–474.

Ball, D. L., Thames, L., & Phelps, G. (2008). Content knowledge for teaching: What makes it special? *Journal of Teacher Education, 59*(5), 389–407.

Barth, Ch., & Henninger, M. (2012). Fostering the diagnostic competence of teachers with multimedia training – A promising approach? In I. Deliyannis (Ed.), *Interactive multimedia* (pp. 49–66). Rijeka, Croatia: InTech.

Bobis, J., Clarke, B., Clarke, D., Thomas, G., Wright, R. B., Young-Loveridge, J., & Gould, P. (2005). Supporting teachers in the development of young children's mathematical thinking. *Mathematics Education Research Journal, 16*(3), 27–57.

Borko, H., Jacobs, J., & Koellner, K. (2010). Contemporary approaches to teacher professional development. In P. Peterson, E. Baker, & B. McGaw (Eds.), *International encyclopedia of education* (Vol. 7, pp. 548–556). Oxford, England: Elsevier.

Bräuning, K., & Steinbring, H. (2011). Communicative characteristics of teachers' mathematical talk with children: From knowledge transfer to knowledge investigation. *ZDM Mathematics Education, 43*, 927–939.

Busch, J., Barzel, B., & Leuders, T. (2015). Promoting secondary teachers' diagnostic competence with respect to functions: Development of a scalable unit in continuous professional development. *ZDM Mathematics Education, 47*, 53–64.

Clarke, D. (2013). Understanding, assessing and developing children's mathematical thinking: Task-based interviews as powerful tools for teacher professional learning. In A. M. Lindmeier & A. Heinze (Eds.), *Proceedings of the 37th Conference of the International Group for the Psychology of Mathematics Education* (Vol. 1, pp. 17–30). Kiel, Germany: PME.

Clarke, D., Clarke, B., & Roche, A. (2011). Building teachers' expertise in understanding, assessing and developing children's mathematical thinking: The power of task-based, one-to-one assessment interviews. *ZDM Mathematics Education*, *43*, 901–913.

Clarke, D., & Wilson, L. (1994). Valuing what we see. *The Mathematics Teacher*, *87*(7), 542–545.

Corbin, J., & Strauss, A. L. (2008). *Basics of Qualitative Research: Grounded Theory Procedures and Techniques* (3rd Ed.). Thousand Oaks, CA: Sage Publications.

Demaray, M. K., & Elliott, S. N. (1998). Teachers' judgements of students' functioning: A comparison of actual and predicted performances. *School Psychology Quarterly: The Official Journal of the Division of School Psychology/APA*, *13*, 8–24.

Fischer, A., & Sjuts, J. (2012). Entwicklung von Diagnose- und Förderkompetenz in Mathematik – ein Modellprojekt zur Verzahnung der Lehrerausbildungsphasen. In M. Ludwig & M. Kleine (Eds.), *Beiträge zum Mathematikunterricht* (pp. 253–256). Münster, Germany: WTM.

Ginsburg, H. & Opper, S. (1998). *Piagets Theorie der geistigen Entwicklung*. Stuttgart: Klett-Cotta, 8. Auflage.

Ginsburg, H. P. (2009). The challenge of formative assessment in mathematics education: Children' s minds, teachers' minds. *Human Development*, *52*, 109–128.

Helmke, A., & Schrader, F.-W. (1987). Interactional effects of instructional ability and teacher judgement accuracy on achievement. *Teaching and Teacher Education*, *3*(2), 91–98.

Hirt, U., & Wälti, B. (2008). *Lernumgebungen im Mathematikunterricht: Natürlich differenzieren für Rechenschwache bis Hochbegabte*. Seelze, Germany: Klett/Kallmeyer.

Hunting, H. (1997). Clinical interview methods in mathematics education research and practice. *Journal of Mathematics Behaviour*, *16*(2), 145–165.

Jungwirth, H., Steinbring, H., Voigt, J., & Wollring, B. (2001). Interpretative classroom research in teacher education. In H. G. Weigand et al. (Eds.), *Developments in mathematics education in Germany* (pp. 49–56). Hildesheim, Germany: Franzbecker.

Jacobs, V. R, Lamb, L. L. C., & Philipp, R. A. (2010). Professional noticing of children's mathematical thinking. *Journal for Research in Mathematics Education*, *41*(2), 169–202.

Kilic, H. (2011). The Nature of Preservice Teachers' Pedagogical Content Knowledge. In: *Proceedings of CERME 7*, 2690–2696. Rzeszow, Poland: CERME.

Klug, J. (2011). *Modelling and training a new concept of teachers' diagnostic competence*. Darmstadt, Germany: TU Darmstadt.

Klug, J., Bruder, S., Kelava, A., Spiel, C., & Schmitz, B. (2013). Diagnostic competence of teachers: A process model that accounts for diagnosing learning behavior tested by means of a case scenario. *Teaching and Teacher Education*, *30*, 38–46.

Lampert, M., & Ball, D. L. (1998). *Teaching, multimedia, and mathematics investigation of real practice*. New York, NY: Teachers College Press.

Lorenz, C., & Artelt, C. (2009). Fachspezifität und Stabilität diagnostischer Kompetenz von Grundschullehrkräften in den Fächern Deutsch und Mathematik. *Zeitschrift für Pädagogische Psychologie*, *23*(3–4), 211–222.

McDonough, A., Clarke, B., & Clarke, D. M. (2002). Understanding, assessing and developing children's mathematical thinking: The power of a one-to-one interview for pre-service teachers in providing insights into appropriate pedagogical practices. *International Journal of Educational Research*, *37*(11), 211–226.

McIntosh, A., & Quadling, D. (1975). Arithmogons. *Mathematics Teaching*, *70*, 18–23.

Moyer, P. S., & Milewicz, E. (2002). Learning to question: Categories of questioning used by preservice teachers during diagnostic mathematics interviews. *Journal of Mathematics Teacher Education*, *5*, 293–315.

Norton, A., McCloskey, A., & Hudson, R.-A. (2011). Prediction assessments: Using video-based predictions to assess prospective teachers' knowledge of students' mathematical thinking. *Journal of Mathematical Teacher Education*, *14*, 305–325.

Peter-Koop, A. (2006). Die Vermittlung mathematikdiagnostischer Kompetenzen in der universitären Grundschullehrerausbildung. In A. H. Hilligus & H.-D. Rinkens (Eds.), *Standards und Kompetenzen – neue Qualität in der Lehrerausbildung?* (pp. 459–467). Berlin, Germany: LIT.

Peter-Koop, A., & Wollring, B. (2001). Student teacher participation in interpretative classroom research. *Mathematics Teacher Education and Development*, *3*, 4–15.

Peter-Koop, A., Wollring, B., Spindeler, B., & Grüßing, M. (2007). *Elementarmathematisches Basisinterview*. Offenburg, German: Mildenberger.

Philipp, K., & Leuders, T. (2014). Diagnostic competences of mathematics teachers – Processes and resources. In C. Nicol, S. Oesterle, P. Liljedahl & D. Allan (Eds.), *Proceedings of the Joint Conference of PME 38 and PME-NA 36* (Vol. 4, pp. 425–432). Vancouver, Canada: PME.

Prediger, S. (2010). How to develop mathematics for teaching and for understanding. *Journal of Mathematics Teacher Education, 13*(1), 73–92.

Reinhold, S. (2014). Qualitative facets of prospective elementary teachers' diagnostic competence: Micro-processes in one-on-one diagnostic interviews. In C. Nicol, S. Oesterle, P. Liljedahl & D. Allan (Eds.), *Proceedings of the Joint Conference of PME 38 and PME- NA 36* (Vol. 5, pp. 41–48). Vancouver, Canada: PME.

Reinhold, S. (2015). Prospective elementary teachers' diagnostic proceeding in one-on-one diagnostic interviews: Facets of data collection and attention. In K. Beswick, T. Muir, & J. Wells (Eds.), *Proceedings of the 39th Conference of the International Group for the Psychology on Mathematics Education* (Vol. 4, pp. 73–80). Hobart, Australia: PME.

Schack, E. O., Fisher, M. H., Thomas, J. N., Eisenhardt, S., Tassell, J., & Yoder, M. (2013). Prospective elementary school teachers' professional noticing of children's early numeracy. *Journal of Mathematics Teacher Education, 16*, 379–397.

Scherer, P., & Moser Opitz, E. (2010). *Fördern im Mathematikunterricht der Primarstufe*. Heidelberg, Germany: Spektrum Akademischer Verlag.

Shulman, L. (1986). Those who understand: Knowledge growth in teaching. *Educational Researcher*, *15*(2), 4–14.

Sleep, L., & Boerst, T. A. (2012). Preparing beginning teachers to elicit and interpret students' mathematical thinking. *Teaching and Teacher Education*, *28*, 1038–1048.

Streit, Ch., & Royar, Th. (2012). Förderung der diagnostischen Kompetenz angehender Lehrpersonen in der Vorschul- und Primarstufe. In M. Ludwig & M. Kleine (Hrsg.), *Beiträge zum Mathematikunterricht 2012* (S. 849–852). Münster: WTM.

Südkamp, A., Kaiser, J., & Möller, J. (2012). Accuracy of Teachers' Judgements of Students' Academic Achievement: *A Meta-Analysis. Journal of Educational Psychology*, 104, 743–752.

Tirosh, D. (2000). Enhancing prospective teachers' knowledge of children's conceptions: The case of division of fractions. *Journal for Research in Mathematics Education, 31*, 5–25.

Wang, M. C. (1992). *Adaptive teaching strategies: Building on diversity*. Baltimore, MD: Brookes.

Whitenack, J. W., Knipping, N., Novinger, S., Coutts, L., & Stanfifer, S. (2000). Teacher' mini-case studies of children's mathematics. *Journal of Mathematics Teacher Education*, *3*, 101–123.

Wilson, P. H., Mojita, G. F., & Confrey, J. (2013). Learning trajectories in teacher education: Supporting teachers' understandings of students' mathematical thinking. *The Journal of Mathematical Behavior*, *32*, 103–121.

Wittmann, E. C. (1981). Unterrichtsbeispiele als integrierender Kern der Mathematikdidaktik. *Journal für Mathematik-Didaktik*, *3*(1), 3–20.

Wollring, B. (1999). Mathematikdidaktik zwischen Diagnostik und Design. In C. Selter & G. Walther (Hrsg.), *Mathematikdidaktik als design science* (S. 270–276). Leipzig u. a.: Klett.

Part II
Developing Diagnostic Competence and Applying Diagnostic Competence in the Classroom: Models for Teacher Training

Developing Diagnostic Competence Through Professional Learning Communities

Karin Brodie, Jeanette Marchant, Nicholas Molefe, and Tinoda Chimhande

In many countries, it is becoming increasingly common for teachers to analyse data from learners'[1] tests and classroom work in order to improve their practice in response to what learners need to learn. In order to use data well, teachers need to develop diagnostic competence, which has been defined as the ability to respond in a didactically sensitive manner to learners' mathematical productions. In this chapter we look at the extent to which mathematics teachers enact elements of diagnostic competence in professional learning communities and their classroom practice. We analyse data from one professional learning community over a two-year period and show that there were features of diagnostic competence in the teachers' conversations and that three of four teachers shifted their diagnostic competence in practice.

1 Introduction

In many countries it is becoming increasingly common for teachers to analyse data from learners' tests and classroom work in order to improve their practice in response to what learners need to learn. This signals a welcome shift in the use of data 'as something that informs teaching and learning, rather than as a reflection of the capability of individual students and to be used for sorting, labeling and credentialing' (Timperley, 2009, p. 21). Teachers can use data to access their learners'

[1] In South Africa, we use the word learners rather than pupils or students to refer to learners at school.

K. Brodie (✉) • J. Marchant • N. Molefe • T. Chimhande
School of Education, University of the Witwatersrand, Johannesburg, South Africa
e-mail: Karin.Brodie@wits.ac.za

T. Leuders et al. (eds.), *Diagnostic Competence of Mathematics Teachers*,
Mathematics Teacher Education 11, DOI 10.1007/978-3-319-66327-2_8

understanding and thinking, and as a support for planning and reflecting on lessons and interacting with learners in the classroom.

In order to use data well, teachers need to develop diagnostic competence (Prediger, 2010). Prediger defines diagnostic competence in mathematics as the ability to respond in a didactically sensitive manner to learners' mathematical productions in order to understand the reasoning behind the learners' thinking, what may be problematic in the learners' thinking and how the teacher might work to reconcile the learners' thinking with the correct mathematical ideas. Becoming diagnostically competent requires substantial work by teachers, and as teacher educators, one of our roles is to support teachers in developing the expertise to do this work.

In this chapter, we draw on data from the Data Informed Practice Improvement Project (DIPIP), an in-service teacher development project in Johannesburg, in which we worked with teachers to develop their responsiveness to learner errors in mathematics. We organized the programme so that teachers worked in professional learning communities, on a cycle of activities that aimed to deepen their knowledge and understanding of learner errors, and through learner errors, their knowledge of teaching and learning mathematics more generally.

In this chapter, we focus on one professional learning community in the project – looking at the teachers' talk in their community and the teachers' practices in their classrooms in relation to their responsiveness to learner errors. We answer the following research question: Do the teachers enact elements of diagnostic competence in the conversations and their practices?

In what follows, we discuss the theoretical background to the study in terms of teacher knowledge, teachers' practices and professional learning communities. Thereafter, we discuss the DIPIP project in more detail, the methods of analysis used for this chapter and then we present our analyses of teachers' responsiveness to learner errors and diagnostic competence in their community conversations and their lessons.

2 Teacher Knowledge

For Prediger (2010), diagnostic competence of learner errors forms part of mathematical knowledge for teaching (Ball, Thames, & Phelps, 2008). Prediger elaborates four key constituents of diagnostic competence. First, the teacher needs to have an *interest* in and be alert to learner errors. Second, this interest needs to be combined with an *interpretive attitude*, which means that she or he needs to strive to view the error from the learner's perspective, understanding the learner's reasoning underlying the error. Such an attitude is important in order to avoid deficit perspectives on learner errors. Third, the teacher needs *general knowledge about learning processes*, in order to think about what learning has produced the error. Fourth, the teacher needs *specific knowledge in the mathematical domain* being taught, so that she or he can analyse the error from a mathematical perspective, which will help her or him to think about how to respond appropriately. Knowledge in the mathematical domain includes knowledge of the correct mathematics as well as the range of meanings that students might develop for mathematical ideas, including misconceptions.

Prediger's constituents of diagnostic competence align with key elements of content knowledge (CK) and pedagogical content knowledge (PCK). Shulman (1987) described CK as conceptual knowledge of the subject, including the organization of the knowledge, the structure of the discipline and why particular ideas are seen as valid in the discipline. He defined PCK as 'the capacity of a teacher to transform the CK he or she possesses into forms that are pedagogically powerful' (Shulman, 1987, p. 15). Since Shulman developed these concepts, much work has been done in elaborating PCK. The most well-known of these is the work of Ball and her colleagues (Ball et al., 2008) who define PCK as comprising knowledge of content and teaching (KCT), knowledge of content and students (KCS) and knowledge of content and curriculum (KCC).

Based on the literature on teacher knowledge in Science Education, Park (2007) presented a model of five integrated components of pedagogical content knowledge: orientation to teaching, knowledge of curriculum, knowledge of assessment, knowledge of students' understanding and knowledge of instructional strategies. The latter two components have been most useful for analysing the data in the DIPIP project. Knowledge of students' understanding (KSU) includes knowledge of students' common misconceptions; in particular, topics as well the difficulties they might have in understanding particular concepts. Knowledge of instructional strategies (KISR) includes general approaches to instruction: inquiry -oriented strategies and topic-specific approaches; for example, what are good ways to teach equations in order to differentiate them from expressions. KSU corresponds with Ball et al.'s notion of KCS while KISR corresponds with KCT and some elements of KCS and KCC.

So CK and PCK, in particular KSU and KISR, form important parts of teachers' diagnostic competence. Helping teachers to become more diagnostically competent requires building these aspects of their knowledge. A key issue in the study of teacher knowledge is how this knowledge is best learned and how do we ascertain whether teachers are developing this knowledge. Much of the literature suggests that CK is best developed outside of the classroom in professional development (PD) programmes, while PCK is best developed in the classroom. We take a different view, arguing that teachers can and should learn both CK and PCK at both sites: their classrooms and their PD programmes. PD programmes work best when there is a two-way interaction between practice and the programme: teachers take what they learn into the classroom and bring their experiences in the classroom to the PD sessions (Kazemi & Hubbard, 2008; Putnam & Borko, 1997), reflecting on CK and PCK together in this process. So while knowledge is important in developing diagnostic competence, it is just as important for teachers to be able to use this knowledge in practice.

3 Teacher Practice

The DIPIP project did not have a particular view of the forms of practice that we wanted teachers to develop; for example, we were agnostic as to whether whole class teaching or group work is preferable as a general principle. What we wanted to develop with teachers was the competence to embrace and engage with learners' errors. We developed this main focus based on what we learned from the literature on errors and misconceptions in mathematics education.

We define learner errors as systematic, persistent and pervasive mistakes performed by learners across a range of contexts (Brodie, 2013; Nesher, 1987). Since errors are systematic and persistent, they are not necessarily responsive to easy correction or re-explanation of concepts (Nesher, 1987; Smith, DiSessa, & Roschelle, 1993). Errors are the performance of misconceptions: each set of errors can give clues to learners' underlying misconceptions that need to be transformed for new mathematical understanding. Errors therefore create possibilities for teachers to access learners' mathematical thinking (Borasi, 1994). We aimed to support teachers to delve more deeply into learners' thinking; to understand what was both valid and not valid in learners' reasoning; and to build on the learners' valid reasoning to shift their misconceptions (Brodie, 2013). This expertise is key to diagnostic competence in practice. It should be clear that we view misconceptions as positive and necessary steps in the development of correct mathematical knowledge and the teachers' diagnostic role is to identify and engage with errors and misconceptions.[2]

In order to engage with learner errors in practice, teachers need a range of in-classroom expertise, drawing on and contributing to their CK and PCK. We drew our understanding of this expertise from the work on mathematical reasoning by Ball and colleagues (Ball & Bass, 2000, 2003; Ball et al., 2008; Hill et al., 2008). A key practice is to develop learners' participation in the classroom, particularly in relation to meaning making and reasoning. Learners can explain how or why certain mathematical ideas do or do not make sense to them; can ask questions of their peers or the teacher; and can come to understand that mathematics is an activity that is as much about reasoning, making sense and communication as it is about getting right answers. Developing learners' participation requires that teachers work with learners' mathematical meanings and put learners' ideas in conversation with each other and with the official mathematical knowledge.

[2] We note that misconceptions can be masked by correct answers, i.e., errors are not the only routes into learners' misconceptions. However, they are important routes and as teachers' diagnostic competence with errors develops, they also begin to see different ways of working with correct answers.

4 Professional Learning Communities

We chose the mode of our PD programme to be professional learning communities (PLCs) based in schools. A PLC is a group of teachers working together in a sustained manner, inquiring collectively into their practices and their learners' learning, with the aim of developing collective, sustainable shifts in practice and improved learner achievement and learning (Katz, Earl, & Ben Jaafar, 2009; Stoll, Bolam, McMahon, Wallace, & Thomas, 2006; Stoll & Louis, 2008). The idea behind professional learning communities is that teachers in the same or similar contexts work together to understand their practice and to develop knowledge and practice strongly related to their contexts. Also, the fact that teachers in the same or nearby schools work together means that the discussions can take place both formally, in the PLC meetings, and less formally because they can continue during the regular school day. In this way PLCs become part of the intellectual life of the school (Stoll et al., 2006).

There are a number of key characteristics of successful professional learning communities: they have a challenging focus; support rigorous enquiry; create productive relationships through trust; and support collaboration for the benefit of teachers and learners (Katz et al., 2009). Safety and challenge, which are mutually supportive, are key elements of successful communities. Teachers need to be safe and trusting enough of their colleagues to admit to their own and their learners' weaknesses. However, too much safety can be unproductive and there needs to be enough challenge to sustain rigorous enquiry into teaching and learning. Successful learning communities challenge their members to reconsider taken-for-granted assumptions; to make tacit knowledge explicit; and to build collective responses to shared problems (Brodie & Shalem, 2011; Katz et al., 2009).

5 The DIPIP Project

DIPIP was a long-term project (2011–2014), based on research that shows that short-term, fragmented seminars and workshops do not work for sustainable professional development (Borko, 2004; Brodie & Shalem, 2011).[3] Teachers were supported to participate in a sequence of developmental activities in which they analysed learners' errors in different teaching contexts: test analysis; learner interviews; curriculum mapping; choosing 'leverage' concepts; readings and discussion; planning lessons; and teaching, videotaping and reflecting on lessons.

The tests that were analysed were international tests, national tests and teacher-set tests, depending on the needs and interest of the community. The test analysis provided an overview of strengths and weaknesses in learners' mathematical

[3]There were three phases to the project and this chapter reports on phase three. The team that conceptualized and implemented phase three consisted of the first and the third author of this chapter and two other graduate students, referred to here as the DIPIP team. The second and fourth authors were not part of the original team but have been involved in analysing the data.

knowledge in a particular school or class. Based on the test analysis, teachers chose learners who had made interesting errors that they wanted to understand more deeply and interviewed these learners. They then took the results of these two analyses and mapped them against the curriculum, working out where the key concepts were taught and what curricular issues might have contributed to the errors (Brodie, Shalem, Sapire, & Manson, 2010). Based on these three activities, teachers chose a leverage concept, which is a concept that underlies many of the errors that learners made in a topic. Two examples of leverage concepts were the equal sign and the differences between equations, expressions and formulae.

Once a concept was chosen, the DIPIP facilitator (see below) found literature on that concept, including learner errors in the concept. The community read and discussed these chapters and drew on these discussions to plan lessons together. The lessons aimed to surface learner errors in the topic and to find ways to engage them. These lessons were taught and videotaped and the community then reflected on episodes of each teacher's lessons in order to understand their strengths and challenges in dealing with learner errors in class.

The DIPIP project design took a particular view of the development of teacher knowledge and practice: that teachers tend to be most focused on what they do every day in their classrooms, that is, their practice and their PCK. Therefore, the best way of developing teachers' knowledge in an integrated way is to start with their practice and PCK and to develop CK in relation to these (Brodie & Sanni, 2014). This is different from many PD programmes, in South Africa in particular, which start with CK and then move on to PCK. Brodie (2013) argues 'one of the key principles of the DIPIP project is that in coming to understand learner needs, teachers can come to understand their own learning needs: what mathematics they need to learn and how to use this new knowledge to improve their practice' (p. 15). Evidence from the project is beginning to show that through working with learner errors, teachers do access their practice, pedagogical content knowledge and content knowledge (Brodie, 2013, 2014; Chimhande & Brodie, 2016; Marchant & Brodie, 2016).

6 Methods

Twelve schools participated in the project, with six schools participating for three or four years. These six schools were organized into three communities, one community per school in three schools and the fourth community, which is the subject of this chapter, was made up of three schools within close proximity to each other. Over the four years about 50 teachers participated in the project, with consistent participation from 22 teachers in the six schools. Participation was entirely voluntary for the school and the teachers within the school. Each community had a facilitator. In the first two years of the project, the facilitators were members of the DIPIP team (university-based), while in the last two years, they were teachers (school-based) who had participated in the PLCs and were trained by the DIPIP team.

The community that we focus on in this chapter participated in all four years of the project, from 2011–2014. We have chosen the years 2013–2014 for analysis in this chapter. This community consisted of three schools very close to each other located in Soweto, an urban township[4] in Johannesburg. One was a junior secondary school (grades 7–9), one was a secondary school (grades 8–12) and the third was a senior secondary school (grades 10–12). The junior secondary school is a feeder school for the senior secondary school. Each teacher chose one class to work with in each year of the project, either grade 8, 9 or 10. In total, seven teachers participated in this community over the four years and in 2013 there were four teachers who participated consistently and one sporadically, while in 2014, the same four teachers participated. Table 1 gives biographic details of these four teachers.

The schools were poorly resourced, with only the very basic equipment of desks, chairs and a chalkboard in each classroom. The community met at the different schools, sometimes in a classroom or science laboratory and sometimes in the staffroom, which was usually a bare room with chairs and desks for the teachers. The communities met once a week for two hours after school time during school terms, which was a big time commitment to the work. The school-based facilitator stayed for an additional half-hour debriefing with a DIPIP team member (previously a university-based facilitator) after every second meeting in 2013 and attended two-hour meetings with the facilitators from the other communities and the DIPIP team on Fridays after school once a month in 2013 and 2014.

In this chapter, we present analyses of two sets of data, the PLC meetings in 2013 and classroom lessons in 2013–2014. These analyses help to answer our research question: Do the teachers enact elements of diagnostic competence in the conversations and their practices?

We videotaped or audiotaped 21 PLC meetings during the year, of which 17 were selected for analysis. The four that were not analysed consisted of two error-capturing sessions, where there was not much discussion, a reflection on a meeting with the other schools and a meeting in which the teachers planned for the following year. Of the 17 lessons, two were on test analysis, two on learner interviews, five on lesson planning and eight on lesson reflection. The other three activities (curriculum mapping, choosing 'leverage' concepts and readings and discussion) were not

Table 1 Teachers[a]

	Role in PLC	Teaching experience (years)	Grades taught in 2013 and 2014
Chamu	Facilitator	18	10, 11
Mapula	Participant	30	8, 9
Funeka	Participant	7	10
Khumo	Participant	20	7, 8

[a]Names are pseudonyms

[4]Townships were established under apartheid as segregated living areas for black South Africans. They remain largely inhabited by black people, and township schools usually have black learners and teachers and are poorly resourced in relation to suburban schools, which are more diverse.

engaged in by this community in 2013 - the community chose the leverage concept during the lesson planning sessions.

Each PLC session was divided into conversation units, which are defined as a period of continuous time with discussion of one topic or one knowledge type (CK or PCK) in the conversation. Each conversation unit was then coded according to activity, knowledge and level.[5] Seven activities were coded: four substantive activities – test analysis, learner interviews, lesson planning and lesson reflection – and three others: set-up, closure or off-topic. The latter three activities were not coded any further and not used in the analysis of data. It is of interest to note that setup, closure and off-topic conversations occupied only 3% of the total conversation time, meaning that the teachers spent 97% of the conversation time discussing PCK and CK.

Conversation units involving the four substantive activities were coded by means of rubrics (see Appendix). The rubrics were developed for three areas of teacher knowledge: CK, PCK (KSU) and PCK (KISR), with the latter two being modifications of Park, Jang, Chen, and Jung's (2011) frameworks for PLCs. Each rubric had four levels: Limited, Basic, Developing and Exemplary. The CK rubric was not divided further but indicators were given for each level. The PCK (KSU) rubric had two categories: identify errors and reasoning behind errors; and identify what makes a topic or concept difficult. Indicators were given for each category, for each level. The PCK (KISR) rubric had four categories: identify and discuss teaching strategies to accommodate learner errors; discuss rationales for teaching strategies in relation to learner errors; discuss how to probe learner reasoning behind errors; and discuss modification of instructional strategies (see Appendix for rubrics).

The coding was done using a software programme called Studiocode™, which allows the video and audio recordings to be plotted along a timeline. The recordings were divided into conversation units and coded as described above. The coding was done by the second author of this chapter, who did both intra- and inter-rater reliability checks on her coding. In the intra-rater reliability checks, she checked the consistency of her own coding over time by re-coding three conversations two weeks after the initial coding and then comparing the results. In the inter-rater reliability checks, the other authors of this chapter coded some sessions and checked with the coder. Both inter- and intra-rater reliability were found to be above 95%. Once all 17 professional learning community meetings had been coded, the Studiocode™ programme was used to generate the matrices and code reports, which form the basis of the data analysis in this chapter.

We also analysed teachers' lessons for 2013–2014. Between two and four lessons were videotaped for each teacher in April or May and August and October of 2013 and in March or May and October of 2014, with a total of 44 lessons analysed.

The lessons were analysed with the Mathematical Quality of Instruction (MQI) coding tool (Hill et al., 2008). MQI refers to 'a composite of several dimensions that characterize the rigor and richness of the mathematics lesson, including the presence

[5] We do not discuss level in detail in this chapter, although we do refer to it.

or absence of mathematical errors, mathematical explanation and justification, mathematical representation, and related observables' (Hill et al., 2008). Using this tool entails coding teachers' lessons on five key dimensions that cumulatively contribute to the quality of the lesson. The dimensions are: Mode of Instruction; Richness of Mathematics; Working with Students and Mathematics; Errors and Imprecision; and Student Participation in Meaning-Making and Reasoning. Each dimension has sub-dimensions that elaborate the dimension. Coding of each sub-dimension was done by judging the extent to which each activity or phenomenon was observed in each ten-minute interval. We coded: 'Low or None' (1); 'Mid or Some' (2); and 'High or Most' (3).

This coding was done by a graduate student who was trained by the third author of this chapter, who had already coded the first two years of these teachers' lessons. The process of coding entailed: discussing the MQI instrument; coding some lessons together and discussing agreement or disagreement after each episode, and coding some lessons separately and calculating inter-rater reliability, which came to 75%. After the lessons were coded, the third author reviewed the coding and disagreements were discussed and resolved by consensus. We then averaged the scores for each dimension over the episodes for each year (Koellner & Jacobs, 2015) and drew up tables and graphs, which show the shifts from 2013 to 2014.

We note that neither of our instruments were developed in relation to the specific notion of diagnostic competence as discussed above. However, they can show us how teachers spoke and enacted aspects of diagnostic competence in the PLC and their classrooms.

7 Examples: PLC Conversations

Before we present our analysis, we give two examples of PLC conversations: one CK conversation and one PCK conversation. We do this to exemplify our coding and give some indication of what counts as examples of the indicators in our rubrics.

The first example consists of extracts from a CK conversation unit, which occurred during a lesson reflection session. The teachers were talking about a learner error in relation to the statement: two in every five teachers owned cars. This was interpreted as the ratio 2:5 and the facilitator initiated a conversation to discuss the conceptual differences between the two statements. It turned out that one of the teachers (Khumo) struggled herself with this idea and the conversation shifted from a conversation about a learner to the teachers' own knowledge, with the group finding different ways to explain the ideas.

Chamu	…now I am saying, this two in every five, is it the same as two is to five?
Funeka	Two in every five?
Chamu	Is it the same as two is to five or, so, it means out of those five, how many don't have cars? If I may ask it that way, to say, two in every five have got cars?
Mapula	Three don't have cars.

Chamu	How many in every five don't have cars?
Mapula	Three.
Chamu	So what is the ratio out of that?
Funeka	The total is twenty.
Chamu	Is it two is to three ratio of cars to no cars? Is it two is to three or is it to is to five?
Funeka	Three is to two.
Chamu	Two is to three.

In the above excerpt, Chamu (the facilitator), Funeka and Mapula distinguished 2:3 from 2:5 as the representation of two in every five teachers own cars. However, it became apparent that Khumo was not sure about the difference (Khumo had weaker content knowledge than the other teachers). Chamu used drawings to show that if there were 20 teachers, eight would own cars and 12 would not. This explanation helped Funeka and Mapula to explain to Khumo.

Chamu	Can you two maybe assist her and help her with that?
Khumo	Five turns to twenty.
Funeka	And then you have two is to three.
Khumo	Okay, so three, six and twelve, ne?
Chamu	Twelve don't have.
Funeka	Ja, now if twelve don't have, that means eight have cars.
Mapula	Its eight is to twelve.
Chamu	Okay, reduce that eight is to twelve, it goes down to what? Can you reduce that ratio; eight is to twelve, to its simplest form?
Mapula	Two is to three.

This was coded as a level 3 content knowledge conversation because some new ideas were developed in the conversation. In order to explain to Khumo, the three other teachers found a different representation and worked from the total number of teachers back to the ratio. Their explanation clarified ideas for Funeka and Mapula so that they could explain more carefully for Khumo.

The second example comes from a PCK conversation unit. It occurred during a lesson planning session where the teachers were anticipating the errors that learners might make. The task they wanted to set for the learners was: Find x if $2^{x+1} = 32$

Mandla	Yes, ja, and what is the first thing that learners might do with this equation instead of solving for *x*? When they're trying to solve for *x*, they will try to do what? Personally, I think they'll divide by two both sides. For me, I think they will say two into two, one; two into thirty two.
Khumo	Sixteen.
Mandla	Sixteen.
Funeka	I should think because they will be solving for *x*, they will bring the *x down because the x,* there is the exponent.
Mandla	The *x* plus one.

Funeka	x plus one, they will bring it down to be two x plus one and then start by dividing or expose, transpose one.
Mandla	What they are saying…is two x plus one.
Funeka	Yes, plus one is equals to thirty two.
Mandla	Is equal to thirty two and then expand and transpose what?
Funeka	One.
Mandla	One, so it's going to be negative one and two x.
Funeka	Ja and then they'll solve for x.
Mandla	x will be..?
Mapula	They divided that by itself, fourteen.
Mandla	Fifteen comma five.
Mapula	Which is sixteen.
Funeka	It's fourteen comma five, I think so.
Khumo	Fifteen comma five, ja.
Funeka	It's fifteen comma five? Okay, ja, fifteen comma five.
Mandla	They are ignoring it. They don't even see that.
Funeka	Ja, they don't want to see the exponents. It's difficult for them.

The teachers anticipated two errors that the learners might make. Mandla suggested that the learners might divide both sides by two, while Funeka suggested that learners would ignore the fact that $x + 1$ is an exponent and work with $2x + 1 = 32$ to get the answer 15.5. This conversation unit was coded as PCK-KSU-identify errors. The teachers were clearly identifying the potential errors that the learners might make with this task. It was coded at level 2 because while the teachers identified possible errors, they did not discuss the reasoning behind the potential errors.

8 Analysis: PLC Conversations

Our analysis focuses on our question: Do the teachers enact elements of diagnostic competence in the conversations and their practices? First, we show the distributions of time spent in CK and PCK conversations, as well as in KSU and KISR conversations, in relation to the different activities that the teachers engaged in. We argue that the teachers did enact aspects of diagnostic competence in their conversations in relation to Prediger's (2010) four components of diagnostic competence.

Table 2 shows that the teachers spent 34% of the total conversation time in CK conversations and 66% of the total conversation time in PCK conversations. This finding suggests that an approach that attempts to build diagnostic competence through a focus on PCK can help teachers to build both their CK and their PCK.

Table 3 shows that lesson reflection and lesson planning were the two activities that accounted for the majority of the conversation time, accounting for 43% and 41% of the total conversation time, respectively. The high percentage of conversation

Table 2 Time spent on CK and PCK conversations

Type of conversation	Time[a]	Percentage of total teacher knowledge conversation time
CK	04:47	34
PCK	09:10	66
Total	**13:57**[b]	**100**

[a]hh:mm; all times are rounded to the minute
[b]This number is lower than the total coded time of 14:46 (Table 3) because there was some off-topic conversation time in the total conversation time

Table 3 Percentage of conversation time by activity type

Activity type	Conversation time	Percentage of total conversation time
Error analysis	01:51	13
Learner interviews	00:40	3
Lesson planning	05:45	41
Lesson reflection	06:26	43
Total	**14:46**	**101**

Table 4 CK and PCK conversation time by activity type

Activity type	Percentage of CK conversation time	Percentage of PCK conversation time
Error analysis	11	14
Learner interviews	9	<1
Lesson planning	58	32
Lesson reflection	22	53
Total	**100**	**100**

time devoted to these activities reflects the chosen focus of the PLC for that year – there were more sessions devoted to these activities.

Looking at the breakdown of CK and PCK time by activity type (Table 4), we show that the amount of time the PLC spent talking about CK and PCK is closely related to the activity type. Table 4 shows that lesson planning activities provided teachers with more opportunities for the development of CK; and that lesson reflection activities provided teachers with more opportunities for the development of PCK.

Lesson planning activities elicited the most CK conversations, with 58% of the total CK conversation time occurring during these sessions, even though lesson planning activities accounted for 41% of the total conversation time (Table 2). During lesson planning meetings, teachers attempted the lesson tasks themselves and when they encountered difficulties, they spoke about the content with their colleagues. The second highest amount of CK conversation time – 22% of total CK conversation time – occurred during lesson reflection activities. During lesson reflection meetings, teachers' analyses of learner errors in class provided the stimulus for some CK conversations (see also Brodie, 2014).

Table 5 KSU and KISR in professional learning community conversations

	Percentage of PCK time	Percentage of total conversation time
KSU	42	27
KISR	58	38
Total PCK time	**100**	**66**

There was more PCK conversation during lesson reflection activities than in any other activity type, with 53% of the total PCK conversation time occurring during lesson reflection activities (Table 3), even though lesson reflection activities accounted for 43% of the conversation time (Table 2). This is because in these meetings the teachers focused on their practice and their engagement with learners' errors. A significant amount of time (32%) of PCK conversation time also occurred in the lesson planning sessions. In these cases, the teachers anticipated learner errors and spoke about teaching strategies to engage with the errors.

An analysis of time spent on the different types of PCK conversation revealed that less PCK conversation time was spent on KSU than KISR (see Table 5). A total of 42% of the PCK time and 27% of the total conversation time was spent on KSU conversation; and 58% of the PCK conversation time and 38% of the total conversation time was spent on KISR conversations.

This was an unexpected finding, given the fact that DIPIP prioritizes understanding learner thinking (KSU) ahead of practice (KISR). The fact that there was more KISR conversation time than KSU conversation time can be explained, at least in part, by the fact that most conversations around errors (the kernel of KSU conversations) quickly led to KISR conversations which centred on instructional strategies for dealing with the errors. In other words, KSU conversations often triggered KISR conversations. This finding can inform future iterations of the DIPIP project where facilitators' attention can be drawn to guard against the tendency to move too quickly into discussing practice before completing dicussions of the reasoning underlying the errors.

Within the category of KSU, more time was spent identifying errors and learners' reasoning behind the errors than on discussing what makes a topic or concept difficult. Table 6 shows that the majority of KSU conversation time (88%) was spent identifying errors and the reasoning behind errors. This suggests that during the KSU conversations, teachers were developing their diagnostic competence in relation to learner errors.

Table 7 shows that the majority of KISR conversation time (71%) was spent discussing teaching strategies to accommodate errors and misconceptions. This makes sense, given that many of the KISR conversations about teaching strategies followed on from KSU conversations about learner errors.

Although we did not develop our analysis in relation to Prediger's (2010) four constituents of diagnostic competence, our results do speak to teachers' opportunities for developing this competence. Teachers clearly had an interest, prompted by the DIPIP project activities, in learners' errors and spent a substantial amount of time talking about them. Although we have not shown it here, the majority of PCK

Table 6 Time spent on each category of KSU conversation

Activity type	Percentage of KSU time
Errors and reasoning behind errors	88
What makes topic/concept difficult	12
Total	**100**

Table 7 Time spent on each category of KISR conversation

Activity type	Percentage of KISR time
Teaching strategies to accommodate errors	71
Rationale for teaching strategies in relation to learner errors	17
Probing learner understanding	3
Modification of instructional strategies based on learner errors	9
Total	**100**

conversations took place either at level 2 (45%) or level 3 (50%) (Marchant, 2015) suggesting that for about half of the time, teachers showed an interpretive attitude towards learner errors and their engagement with learner errors. In relation to specific mathematical knowledge, we see that teachers spent a third of their time talking about their own mathematical knowledge, and we have shown elsewhere (Marchant, 2015) that about 76% of this talk was at level 3. We did not find significant general knowledge about learning distinct from PCK- if we had, we would have modified the rubric to take account of it. We note that PCK could include both general knowledge and domain specific knowledge. We suspect, based on previous analysis (Brodie, 2014), that the general knowledge might be emphasized more by university-based facilitators than school-based facilitators, but we would need to analyse our data further to substantiate this claim.

We now turn to an analysis of the teachers' lessons, and show where they were able to integrate aspects of diagnostic competence into their teaching.

9 Analysis: Classroom Teaching

Since diagnostic competence is defined as the ability to respond in a didactically sensitive manner to learners mathematical productions, the two dimensions from the MQI instrument most important in relation to diagnostic competence are: working with students[6] and mathematics and student participation in meaning making and reasoning. Working with students and mathematics captures whether teachers

[6]We follow the use of "student" in the MQI instrument (Hill et al., 2008).

can understand and respond to students' mathematical ideas, including questions, claims, explanations, solutions or errors. For a low score (1), there are very few student ideas or errors or the teacher does not respond to learner ideas or errors. For a mid-score (2), the teacher responds mainly procedurally although there may be some brief work with the reasoning behind students' ideas. For a high score (3), the teacher understands, engages with and predicts student responses in a conceptual manner.

Student participation in meaning-making and reasoning focuses on the students' actions and captures the extent to which students are involved in substantively mathematical tasks and the extent to which students participate in and contribute to meaning-making and reasoning. A low score indicates very little student engagement or engagement with low cognitive demand. A mid-score indicates that students engage somewhat procedurally but with some higher levels of conceptual engagement. A high score suggests strong conceptual engagement by students, with them making conjectures, giving generalized explanations and showing reasoning.

Table 8 shows that all four teachers averaged between 1 and 2 on these two dimensions. This is partly because we averaged the scores and partly because the teachers started off from a very low base with predominantly values of '1' in 2011.

Table 8 shows small shifts for each of the teachers within the 1–2 band. Mapula and Khumo improved on both dimensions, with Chamu improving on one. Funeka decreased on both dimensions. The data show that Chamu and Funeka both achieved some high (3) levels, but not enough to lift their averages. This analysis suggest that the teachers were enacting some aspects of diagnostic competence in their

Table 8 Shifts 2013–2014

Mapula	2013	2014
Working with students and mathematics	1.46	1.62
Student participation in meaning making and reasoning	1.40	1.48
Chamu	**2013**	**2014**
Working with students and mathematics	1.15	1.25
Student participation in meaning making and reasoning	1.30	1.28
Khumo	**2013**	**2014**
Working with students and mathematics	1.24	1.31
Student participation in meaning making and reasoning	1.12	1.32
Funeka	**2013**	**2014**
Working with students and mathematics	1.20	1.10
Student participation in meaning making and reasoning	1.34	1.16

Table 9 Shifts 2011–2012

Mapula	2011	2012
Working with students and mathematics	1.30	1.59
Student participation in meaning making and reasoning	1.19	1.42
Khumo	**2011**	**2012**
Working with students and mathematics	1.11	1.15
Student participation in meaning making and reasoning	1.18	1.19
Funeka	**2011**	**2012**
Working with students and mathematics	1.21	1.64
Student participation in meaning making and reasoning	1.28	1.51

classrooms and improved over the two years, although they did not reach the higher level of what might constitute enquiry classrooms. We note that these numbers and shifts are similar to those shown in a project in the United States (Koellner & Jacobs, 2015), suggesting that the influences of PD programmes are modest in these areas even when the programme focuses explicitly on them.

Although the years 2011–2012 are not the focus of this chapter, we do have results from these years for three of the teachers[7] and they are instructive.

Tables 8 and 9 taken together show that Mapula increased steadily from 2011 to 2014, with a slight dip in 2013. Khumo started off the lowest of the three teachers and increased steadily, in very small increments in 2012, but more quickly after that. Funeka increased more than the others from 2011 to 2012, achieving a number of high scores in 2012, but after that declined substantially, ending off lower than she started.[8] These earlier results support our claim that changes are slow and somewhat steady.[9]

In what follows we describe an episode from 2012 where Funeka worked well with learners' reasoning and participation. We have chosen this episode because it illustrates the shift in teaching that was possible, although in this case, not sustained.

The class had a worksheet with a number of statements, which they had to identify as true or false and give reasons. The statement in this episode was $(3 + 1)^2 = 3^2 + 1^2$. One learner was convinced that the statement was true, stating that when the 3 in the bracket is squared it would give the 3^2 and squaring the 1 would give 1^2. Some learners agreed with this learner, while others disagreed and Funeka did not indicate whether she agreed or disagreed. Learners who disagreed explained why

[7] Chamu only joined the project towards the end of 2012 and we only videotaped his lessons in 2013 and 2014.

[8] Funeka struggled with illness during 2013 and 2014 and was often absent from school and from the PLC.

[9] The small changes are also a result of the 3-point scale on the MQI. A larger scale, which differentiates more in the middle band, would probably be more appropriate for the teachers in our study.

the statement was false. Funeka then asked the first learner who said the statement was correct if he was convinced, and the learner agreed in a very low voice, which probably meant that he was not convinced. This episode occurred in the last period of the day and the class ended without resolving the matter.

The following day Funeka started the lesson by revisiting the problem, and again she wrote: $(3 + 1)^2 = 3^2 + 1^2$ on the board. She asked the same learner who said the statement was correct if he had changed his mind, and the learner said he still thought that the statement was correct. Funeka then gave the class the binomial expression $(2a + b)^2$ to work out, which they eventually got correct after a few struggled with the algebraic manipulations. However, this example did not convince the original learner to change his mind. He argued that $(3 + 1)^2$ is not a product, whereas $(2a + b)^2$ is a product. Funeka gave three more learners the chance to come to the board to show how they would work out the two problems. Some showed that $3 + 1 = 4$ and $4^2 = 16$, thus creating a possibility for seeing the first statement as false. Funeka then drew all the learners' ideas together and showed how the two binomials were the same in that both terms had to be multiplied by both terms in each case, and also different in that the first could be calculated numerically. Some learners remained unconvinced.

In this episode we see learners working on a high-level task, which requires seeing similarities and differences between binomial tasks with and without variables. The task anticipated a common learner error, which came up in the test analysis and in the teachers' classrooms, where learners do not distribute in binomial multiplications with no variables. The task also anticipated learners' difficulties in working with numbers only when in an algebraic context. Funeka allowed an incorrect solution to stay on the board because she wanted learners to discuss it and to make their reasoning explicit. She was responsive to learners' solutions, both correct and incorrect and put them into conversation with each other. So she had an orientation to work with learner errors, an interpretive attitude and the mathematical knowledge to support her to work with this challenging aspect of algebra.

10 Conclusions

In this chapter, we have shown the extent to which a group of mathematics teachers developed diagnostic competence in two sites of practice: their classrooms and their PLC, where they discussed their practice on an ongoing basis. We showed that in the PLC conversations, there were features of diagnostic competence, in particular an interpretive attitude to learner errors and discussions of learner errors as part of KSU and PCK. We showed that three of the four teachers shifted in their diagnostic competence in their teaching during this time, although in a somewhat narrow band.

We cannot make explicit links between the PLC conversations and the teachers' lessons because this was not a control group design. We have analyses from other communities where we saw similar kinds of conversations and similar shifts in teaching. We also note that some of the lessons were planned by the teachers together, taking account of what they had learned in the community, while some were not. These lessons did not show significant differences from each other, so we believe that we have tentative support for the claim that the shifts were related to what was learned in the PLC.

We have shown that the key elements of diagnostic competence were present in the PLC conversations and there was some improvement in the teachers' classrooms practices on two dimensions related to diagnostic competence. PLCs provide opportunities for the development of diagnostic competence in teachers' talk and to some extent in their classrooms.

Appendix: Rubrics for Coding Teacher Knowledge

Table 10 Content knowledge rubric

CK				
	1 Limited	2 Basic	3 Developing	4 Exemplary
Content knowledge	Limited or no discussion	Some discussion but without new ideas in the conversation	Substantial discussion with some new ideas developed in the conversation.	Substantial discussion with substantially new ideas developed in the conversation.

Table 11 Knowledge of student understanding rubric

Knowledge of learner understanding with respect to subject matter (KSU)				
	1 Limited	2 Basic	3 Developing	4 Exemplary
1. Identify errors and discuss reasoning behind errors	No consideration of learner prior knowledge or errors	Identify errors or prior knowledge but go beyond errors to discuss reasoning	Identify errors or prior knowledge and discuss explanations for the error that takes some account of learner reasoning behind the error	Identify errors or prior knowledge and discuss fully (possible) learner reasoning behind the error
2. Identify and discuss what makes topic/ concept difficult	Identify general concepts without specifying sub-concepts that are problematic and do not discuss reasons for difficulties	Identify specific concepts but provide broad generic reasons	Identify specific concepts with reasons related to specified prior knowledge of learners or common misconceptions	Provide reasons linking to specific leverage concepts that when not fully understood adds to the difficulty of a concept regarded as difficult

Table 12 Knowledge of instructional strategies rubric

Knowledge of instructional strategies (KISR)				
	1 Limited	2 Basic	3 Developing	4 Exemplary
1. Identify and discuss teaching strategies to accommodate learner errors	Teaching strategies not identified	Teaching strategies identified and discussed in relation to learner errors but do not take into account reasoning behind learner errors.	Significant integration of reasoning behind learner errors into teaching strategies and some discussion of learner involvement in teaching strategies.	Significant integration of reasoning behind learner errors into teaching strategies and substantive discussion of learner involvement in teaching strategies.
2. Discuss rationales for teaching strategies in relation to learner errors	Teaching strategies not identified	Weak rationale for teaching strategies in connection with learner errors	Adequate rationale for teaching strategies in connection with learner errors	Strong rationale for teaching strategies in connection with learner errors.
3. Discuss how to probe learner reasoning behind errors	No discussion	Some discussion of possible probes which may not address reasoning behind learner errors	Good discussion of possible probes, which address some reasoning behind learner errors.	Strong discussion of possible probes which address reasoning behind learner errors substantively.
4. Discuss modification of instructional strategies based on learner errors.	No discussion of changes to instructional strategies.	Some discussion of changes to instructional strategies taking into account learner errors	Good discussion of changes to instructional strategies taking into account reasoning behind learner errors	Strong discussion of changes to instructional strategies taking into account reasoning behind learner errors substantively.

References

Ball, D. L., & Bass, H. (2000). Interweaving context and pedagogy in teaching and learning to teach: Knowing and using mathematics. In J. Boaler (Ed.), *Multiple perspectives on teaching and learning mathematics* (pp. 83–104). Westport, CT: Ablex.

Ball, D. L., & Bass, H. (2003). Making mathematics reasonable in school. In J. Kilpatrick, W. G. Martin, & D. E. Schifter (Eds.), *A research companion to principles and standards for school mathematics* (pp. 27–44). Reston, VA: National Council of Teachers of Mathematics.

Ball, D. L., Thames, M. H., & Phelps, G. (2008). Content knowledge for teaching: What makes it special? *Journal of Teacher Education, 59*(5), 389–407.

Borasi, R. (1994). Capitalizing on errors as "Springboards for Inquiry": A teaching experiment. *Journal for Research in Mathematics Education, 25*(2), 166–208.

Brodie, K. (2013). The power of professional learning communities. *Education as Change*, *17*(1), 5–18.
Brodie, K. (2014). Learning about learner errors in professional learning communities. *Educational Studies in Mathematics*, *85*, 221–239.
Brodie, K., & Sanni, R. (2014). We won't know it since we don't teach it: Interactions between teachers' knowledge and practice. *African Journal for Research in Mathematics, Science and Technology Education*, *18*(2), 188–197.
Brodie, K., & Shalem, Y. (2011). Accountablity conversations: Mathematics teachers learning through challenge and solidarity. *Journal for Mathematics Teacher Education*, *14*, 419–439.
Brodie, K., Shalem, Y., Sapire, I., & Manson, L. (2010). Conversations with the mathematics curriculum: Testing and teacher development. In V. Mudaly (Ed.), *Proceedings of the eighteenth annual meeting of the Southern African Association for Research in Mathematics, Science and Technology Education (SAARMSTE)* (pp. 182–191). Durban: University of KwaZulu-Natal.
Borko, H. (2004). "Professional development and teacher learning: mapping the terrain." Educational Resarcher 33(8), 3–15.
Chimhande, T., & Brodie, K. (2016). Relationships between activity, content and depth of mathematics teachers' talk in a professional learning community. In W. Mwakapenda, T. Sedumedi & M. Makgato (Eds.), *Proceedings of the 24th Annual Conference of the Southern African Association for Research in Mathematics, Science and Technology Education (SAARMSTE).* (pp. 23–35).
Hill, H. C., Blunk, M. L., Charalambous, M. C., Lewis, J. M., Phelps, G., Sleep, L., & Ball, D. L. (2008). Mathematical knowledge for teaching and the mathematical quality of instruction: As eploratory study. *Cognition and Instruction*, *26*, 430–511.
Katz, S., Earl, L., & Ben Jaafar, S. (2009). *Building and connecting learning communities: The power of networks for school improvement*. Thousand Oaks, CA: Corwin.
Kazemi, E., & Hubbard, A. (2008). New directions for the design and study of professional development. *Journal of Teacher Education*, *59*(5), 428–441.
Koellner, K., & Jacobs, J. (2015). Distinguishing models of professional development: The case of an adaptive model's impact on teachers' knowledge, instruction, and student achievement. *Journal of Teacher Education*, *66*(1), 51–67.
Marchant, J. (2015). *The Development of content knowledge and pedagogical content knowledge in a professional learning community* (Unpublished Masters Research Report). Johannesburg: University of the Witwatersrand.
Marchant, J., & Brodie, K. (2016). Content knowledge and pedagogical content knowledge conversations. In W. Mwakapenda, T. Sedumedi & M. Makgato (Eds.), *Proceedings of the 24th Annual Conference of the Southern African Association for Research in Mathematics, Science and Technology Education (SAARMSTE).* (pp. 148–159).
Nesher, P. (1987). Towards an instructional theory: The role of students' misconceptions. *For the Learning of Mathematics*, *7*(3), 33–39.
Park, S. (2007). Teacher efficacy as an affective affiliate of pedagogical content knowledge. *Journal of the Korean Association of Research in Science Education*, *27*(8), 743–754.
Park, S., Jang, J., Chen, Y., & Jung, J. (2011). Is Pedagogical Content Knowledge (PCK) necessary for reformed science teaching?: Evidence from an empirical study. *Research in Science Education*, *41*, 245–260.
Prediger, S. (2010). How to develop mathematics-for-teaching and for understanding: The case of meanings of the equal sign. *Journal for Mathematics Teacher Education*, *13*(1), 73–93.
Putnam, R., & Borko, H. (1997). Teacher learning: Implications of the new view of cognition. In B. J. Biddle, T. L. Good, & I. F. Goodson (Eds.), *The international handbook of teachers and teaching* (pp. 1223–1296). Dordrecht, Netherlands: Kluwer.
Shulman, L. S. (1987). Knowledge and teaching: Foundations of the new reform. *Harvard Educational Review*, *57*(1), 1–22.
Smith, J. P., DiSessa, A. A., & Roschelle, J. (1993). Misconceptions reconceived: A constructivist analysis of knowledge in transition. *The Journal of the Learning Sciences*, *3*(2), 115–163.

Stoll, L., Bolam, R., McMahon, A., Wallace, M., & Thomas, S. (2006). Professional learning communities: A review of the literature. *Journal of Educational Change*, *7*, 221–258.

Stoll, L., & Louis, K. S. (2008). Professional learning communities: Elaborating new approaches. In L. Stoll & K. S. Louis (Eds.), *Professional learning communities: Divergence, depth and dilemmas* (pp. 1–13). Maidenhead, Berkshire: Open University Press and McGraw Hill Education.

Timperley, H. (2009). Using assessment data for improving teaching practice: Collecting, Interpreting and using data to inform teaching. http://research.acer.edu.au/cgi/viewcontent.cgi?article=1036&context=researchconference

Supporting Mathematics Teachers' Diagnostic Competence Through the Use of One-to-One, Task-Based Assessment Interviews

Doug M. Clarke, Anne Roche, and Barbara Clarke

In this chapter, the important role that one-to-one, task-based assessment interviews can play in developing inservice and preservice mathematics teachers' diagnostic competence is presented. We argue that the use of such interviews builds competence through enhancing teachers' knowledge of individual and group understanding of mathematics, including misconceptions and preferred strategies, while providing an understanding of the typical learning paths in various mathematical domains. The use of such interviews also provides a model for teachers' interactions and discussions with children in classrooms, building both pedagogical content knowledge and subject matter knowledge.

1 Introduction

The use of a research-based, one-to-one assessment interview set within a well-designed and supportive professional learning context can make an important contribution to inservice and preservice mathematics teachers' diagnostic competence and thereby their knowledge in action.

This chapter draws substantially on a previous paper by the authors (Clarke, Clarke, & Roche, 2011. Building teachers' expertise in understanding, assessing and developing children's mathematical thinking: the power of task-based, one-to-one interviews. *ZDM Mathematics Education, 43*(6), 901–913.)

D.M. Clarke (✉) • A. Roche
Australian Catholic University, Melbourne, Australia
e-mail: Doug.Clarke@acu.edu.au; Anne.Roche@acu.edu.au

B. Clarke
Monash University, Melbourne, Australia
e-mail: barbara.clarke@monash.edu

T. Leuders et al. (eds.), *Diagnostic Competence of Mathematics Teachers*,
Mathematics Teacher Education 11, DOI 10.1007/978-3-319-66327-2_9

An overview of two research programs in which the one-to-one interview formed a major component is given. Examples are given of the kinds of assessment tasks used and the learning frameworks which underpinned them. The professional learning input which preceded teachers' use of the interviews is outlined. Student performance data on certain tasks are given. Teacher questionnaire and individual and focus group data are drawn upon, where relevant, as well as anecdotes from the research teams. Finally, recent developments are shared involving the use of the Early Numeracy Interview with children with Down syndrome.

With a focus on *student learning* and growth in such learning over time as evidenced from interview data, the research projects were not originally intended to study the direct contribution of the one-to-one interview to developing *teacher diagnostic competence*. However, the data on this which emerged during the projects were compelling.

2 The Use of Clinical Interviews in Mathematics in Australia

In the last 20 years, the inadequacy of a single assessment method administered to students at the end of the teaching of a mathematics topic to provide valid evidence of the understanding of an individual has been widely acknowledged (Ginsburg, 2009). It is increasingly the case that those working at all levels of mathematics education regard the major purpose of assessment as improving instruction and supporting learning (Webb & Romberg, 1992), and this has led to a search for appropriate assessment methods. The limitations and disadvantages of pen and paper tests in gathering high-quality, in-depth data on children's knowledge were well established by Clements and Ellerton (1995). They contrasted the quality of information about Grade 5 and Grade 8 students gained from written tests with that gained through one-to-one interviews. They observed that children may have a strong conceptual knowledge of a topic (revealed in a one-to-one interview) but be unable to demonstrate that during a written assessment.

Clinical interviews have been used for many years in mathematics education research (Ginsburg, Klein, & Starkey, 1998), usually with small numbers of students, and the results not always communicated well to the teaching profession. However, the late 1990s, in Australia and New Zealand, saw the development and use of research-based one-to-one, task-based interviews with large numbers of young students, as a professional tool for teachers of mathematics (Bobis et al., 2005).

3 Two Large-Scale Research and Professional Learning Projects Involving Extensive Use of Assessment Interviews

3.1 The Early Numeracy Research Project

The Early Numeracy Research Project (ENRP), a research and professional development program conducted in Victoria from 1999 to 2001, involved 353 teachers and over 11,000 children aged five to eight years old (Clarke et al., 2011; Clarke et al., 2002).

There were four key components to this research and professional development project:

- The development of a research-based *framework of "growth points"* in young children's mathematical learning (in number, measurement, and geometry);
- The development of a 40-minute, one-to-one, task-based *interview*, used by all teachers to assess aspects of the mathematical knowledge of all children at the beginning and end of the school year;
- Extensive *teacher professional development* at central, regional, and school levels, for teachers, mathematics coordinators, and principals; and
- A study of the *practices of particularly effective teachers.*

It was decided to create a framework of key "growth points" in numeracy learning. Students' movement through these growth points in project schools, as revealed in interview data, could then be tracked over time. The project team studied available research on key "stages" or "levels" in young children's mathematics learning (e.g., Clements, Swaminathan, Hannibal, & Sarama, 1999; Lehrer & Chazan, 1998), as well as frameworks developed by other authors and groups to describe learning. Data relating to growth points and student learning have been reported in a number of publications (see, e.g., Clarke, 2004; Clarke, Clarke, & Cheeseman, 2006).

The one-to-one interview, taking an average of 45 min, followed a fairly tight "script," which indicated precisely which question to ask next, given a particular response to the previous item. It was emphasized to interviewers that the interview was for assessment purposes, not an opportunity to teach the student. The interviews were conducted by the student's regular classroom teacher, following a full day's training on its use and the opportunity to practice the interview process under the supervision of either the school mathematics coordinator (who had received additional training) or a member of the research team.

A range of procedures was developed to maximize consistency in the way in which the interview was administered across the schools. This highlights the dual intent of the interview for building teachers' diagnostic competence and the project data providing the opportunity to make valid and reliable statements about a larger group. The teacher completed a record sheet during each interview, which recorded both students' answers and their stated or observed strategy. There was effectively no time limit on students' responses, although when it became clear that the student had little idea on how to attempt to solve a given problem, the teacher would usually move on.

The interview provided information about growth points achieved by a child in each of nine mathematical domains: four in number (counting, place value, addition and subtraction, multiplication and division); three in measurement (time, length, mass), and two in geometry (properties of shape and visualization and orientation). Although the full text of the ENRP interview involved around 60 tasks (with several sub-tasks in many cases), no child was posed all of these. The interviewer made a decision after each task, according to the script. Given success on a particular task, the interviewer continued with the next task in the domain as far as the child could go with success. Given difficulty with the task, the interviewer either abandoned that section of the interview and moved on to the next domain or moved into a "detour," designed to elaborate more clearly the difficulty a child might be having with a particular content area.

Figure 1 includes some questions from the interview (Department of Education and Training, 2001). These questions focus on identifying the strategies that the child draws upon for multiplication. In the example, text in italics is an instruction to the interviewer, and the words which the interviewer is to say are in normal type. The strategies used were recorded on the interview record sheet.

Since its development, the ENRP interview has been used by teachers and researchers (translated as appropriate) in at least eight other countries.

3.2 Australian Catholic University (ACU) Rational Number Interview

Given the impact on the diagnostic competence of teachers of young children, it was decided to develop a second one-to-one interview for teachers of 9 to 14-year olds. Given the recognized challenges faced by teachers and students with the teaching and learning of fractions and decimals (see, e.g., Behr, Lesh, Post, & Silver, 1983; Steinle & Stacey, 2003), it was decided to make rational numbers the focus of the interview. An important source of assessment tasks was the *Rational Number Project* (Behr & Post, 1992). Student data on key tasks were reported in Clarke, Roche, Mitchell, and Sukenik (2006). As part of this project, this interview was used with 323 students who were completing the last of seven years of primary school (typically 11 and 12 years old).

An important difference between the two projects was that the Rational Number Interview was not embedded in a system-wide professional learning program, but rather in a series of small professional learning programs (with voluntary participants), and the interview was accessed by many teachers online, on the Victorian Education Department website.

Two sample tasks from the ACU Rational Number interview are given in Fig. 2, including the percentage success of Grade 6 students on these tasks at the end of the school year. These are *Ordering a large set of ragged decimals* (Roche, 2005) and *Simple operators* (Clarke et al., 2006).

29. Tennis Balls Task

Put out 1 packet of 3 tennis balls.

a) How many balls would there be in four packets?
b) Tell me how you worked that out.
c) *If the child appears to be counting all, ask:* Could you do that another way, without counting them one by one?

Fig. 1 Tasks from the ENRP interview

Ordering a large set of ragged decimals (9% success)

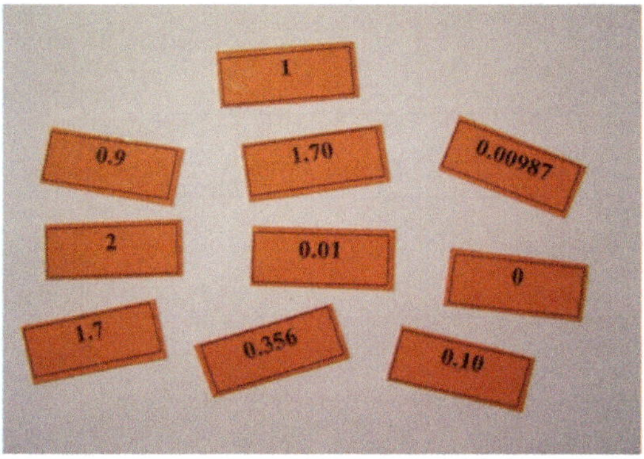

Simple operators

- What is one-half of six? (97% success)
- What is one-fifth of ten? (73% success)
- What is two-thirds of nine? (70% success)
- What is one-third of a half? (18% success)

Fig. 2 Sample tasks from the Australian Catholic University Rational Number Interview.

The first task illustrates the potential of one-to-one interview tasks, as compared to traditional written assessment. The capacity of students to move the cards around has at least three clear benefits. First, the student can place them in particular positions initially, knowing that they are easily changed. Second, the teacher conducting the interview has a window into children's reasoning as they see them move the pieces from place to place, particularly noting decisions related to the placement of "0." Third, the cards make it impractical for students to add zeros to some numbers to equalize the number of decimal places across the set—a common strategy but one which is unhelpful in understanding place value ideas in the longer term (Roche & Clarke, 2004). Students are then more likely to use a conceptual strategy. Such rich information would be difficult to collect from a written assessment.

3.3 Some Information on How the Interviews Were Administered and the Data Collected

Student strategies were recorded in detail on the respective interview record sheet. For example, in the Early Numeracy Interview, for the tasks outlined in Fig. 1, the teacher completed the record sheet, as shown in Fig. 3, recording both the answer given and the strategies used. The emphasis on asking for and recording both answer and strategies is clear recognition that the answer alone is not sufficient, and gives a message to students that their strategies and mathematical thinking are valued (Swan, 2002).

The act of completing the record sheet across the various mathematical domains requires an understanding of the strategies listed (e.g., skip count, modeling, and counting on) and was preceded by extensive teacher professional development on

29. Tennis balls (circle strategy used)
a, b Answer: ________ ☐
- **Skip count**
- **Known fact** ______________________________
- Count, all by 1s
- Other ______________________________

c Answer: ________ ☐
- **Skip count**
- **Known fact** ______________________________
- Count, all by 1s
- Other ______________________________

Fig. 3 An excerpt from the interview record sheet for multiplication.

the use of the record sheet. This is our first example of the kinds of teacher diagnostic competence which are developed prior to and during the use of the interview.

Processes used by the ENRP research team to maximize reliability and validity of interview data have been detailed elsewhere (Horne & Rowley, 2001). Having data on over 36,000 ENRP interviews for the 11,000 students (with many students being interviewed on several occasions) and around 300 for the ACU Rational Number Interview provided previously unavailable high-quality data on student performance. During the ENRP, it became increasingly obvious to the research team that the interview was providing opportunities for the development of diagnostic competence. Evidence emerged of teachers' greater confidence in the use of mathematical language to describe students' strategies, and of their growing sense of the typical learning paths of their students.

3.4 Some Important Similarities and Differences with the Two Projects

As indicated in earlier sections, both interviews have been used extensively by teachers. In the early stages of both projects, the use of the interviews was "controlled" in that those proposing to use the interview were given extensive preparation, and the timing of use was prescribed. For the Early Numeracy Interview, the interviewers were practicing teachers for the most part, while for the Rational Number Interview, all interviewers were part of a trained team of research assistants, all experienced primary teachers.

A common feature of both interviews is that because both interview scripts and materials are now freely available on the Victorian Department of Education and Training website, both interviews are now used in a way and a time of choosing by individuals and groups, and the research team of course has no control over this.

Unlike the Early Numeracy Interview, there was little variation in the use of the Rational Number Interview, with all students being offered all tasks, with few exceptions. There were important reasons for this distinction. Given the large body of research on early mathematics learning, we were able, in advance, to predict that if students were unsuccessful on task A, they would have little chance of success on task B. Our predictions were generally validated in trialing. For the Rational Number interview, however, the research of others and our piloting of interview tasks convinced us that it was much harder to infer the likely success of a student on task B, given their performance on task A. This we put down to the many models, representations, and constructs of fractions and decimals. For example, we found that performance on tasks related to the notion of "fraction as measure" (Clarke et al., 2006) seemed to provide no predictive information for tasks relating to "fraction as operator." Although frameworks have been developed for rational number learning (see, e.g., Fosnot & Dolk, 2002), these were of little help to us in the kind of inferences we were hoping to make.

In the following section, we outline the methodology which led to many of our insights about the potentially important role the interview can play in enhancing teachers' diagnostic competence. Our underlying research question for this part of the research was: What benefits and challenges do teachers report in the use of one-to-one, task based assessment interviews in mathematics?

4 Methodology

Although there was a range of data which is not reported here (e.g., teachers' grouping practices, their planning methods, actual time given to mathematics, and their expectations of student growth), data collection relevant to this chapter took the following forms:

- Teacher Entry questionnaire (February 1999), with 24 items focusing on areas including background information, personal mathematical knowledge, confidence in teaching mathematics, mathematical expectations of students, and areas of their teaching which they sought to improve ($n = 195$).
- Teacher Exit questionnaire (October 2001), involving 21 items focusing on similar areas to the Entry questionnaire, in order to discern changes over time ($n = 221$).
- Teachers' Highlights and Surprises questionnaire (March 1999), where teachers were simply asked "what highlights and surprises were there as a result of conducting the interviews with your students?" ($n = 198$).
- Changes in Teaching questionnaire (October 2001), where teachers were asked to nominate the greatest changes in their teaching and in their students' learning as a result of their involvement in the ENRP ($n = 220$).

All of these data are reported in detail in Clarke et al. (2002). We should stress that there is more robust data for the Early Numeracy Interview than the Rational Number Interview, where teacher responses to the interview have been more anecdotal.

In a separate study, using a two-page questionnaire involving Likert scale items and open response items, 140 *preservice teachers* at Australian Catholic University (ACU) and Monash University were asked to self-report on any growth in diagnostic competence through their use of one-to-one assessment interviews and implications for their future teaching (McDonough, Clarke, & Clarke, 2002). In addition, five ACU preservice teachers were interviewed individually with a particular emphasis on what they had learned in relation to individuals' understanding of mathematics and strategy use. A focus group discussion was also conducted with six other preservice teachers, focusing upon implications for teaching of what they had learned from the interviews. Audiotapes of all interviews and focus groups were transcribed.

"All research is a search for patterns, for consistencies" (Stake, 1995, p. 44). An interpretative perspective (Erickson, 1986) was taken in identifying themes from the unstructured open-response items. All of the responses were read and the main themes identified by two of the researchers, working independently. These were then debated, the terminology clarified, and a set of themes was determined by consensus. One researcher then categorized all data units according to the agreed themes to allow unique categorization within the one theme. This is a form of data reduction in that it groups information into "a smaller number of sets, themes or constructs" (Miles & Huberman, 1994, p. 69). As most teachers had more than one response to the given item, different parts of their response were often coded to different themes. This was generally not a complicated process, as the different responses by a teacher were usually on different themes, often listed as separate dot points. For example, the following statements were given the three different codes as shown.

- *I have a greater understanding of how children learn.* [KHCL: knowledge of how children learn]
- *Working to children's own ability and needs.* [ALN: addressing learning needs]
- *Knowledge of the growth points helps make my planning more detailed.* [GPIP: growth points inform planning]

In the following section, particular tasks, data from teachers, and insights from other researchers are used to build the argument of the power of the interview as an important tool in building teachers' knowledge and expertise in understanding, assessing, and developing children's mathematical thinking.

5 The Role of the Interview and Growth Points in Developing Teachers' Diagnostic Competence in Mathematics Teaching

Sowder (2007) claimed that student thinking could be thought of as an interpretive lens that "helps teachers to think about their students, the mathematics they are learning, the tasks that are appropriate for the learning of that mathematics, and the questions that need to be asked to lead them to better understanding" (p. 164).

There were several common themes in the Changes in Teaching Questionnaire (Clarke, 2008), many of which related directly to enhanced diagnostic competence experienced through the use of the interview. They were, in decreasing order of frequency:

- using growth points to inform planning (63 responses);
- using knowledge of individual understanding to address learning needs (49);
- challenging and extending children and having higher and/or more realistic expectations (33);
- having more confidence in teaching mathematics (28);
- enjoying mathematics more and making mathematics more interesting (27); and
- having a greater knowledge of how children learn (24).

Several of these themes are evident in the following response from a teacher:

> The assessment interview has given focus to my teaching. Constantly at the back of my mind I have the growth points there and I have a clear idea of where I'm heading and can match activities to the needs of the children. But I also try to make it challenging enough to make them stretch.

Our experience is that the claims made here for practicing teachers in relation to diagnostic competence apply also to a large extent to preservice teachers. On a questionnaire which used a Likert scale (McDonough et al., 2002, p. 219), the number of preservice teachers out of 140 *agreeing or strongly agreeing* with a given statement is shown in parentheses:

The interview ...

- gave me new insights into how young children think when doing maths (135);
- gave information that would help me to plan for and teach that child (129);
- gave insights that would help me to plan for and teach all children (92);
- gave me insights into the types of questions to ask young children to assess their understandings and strategy use (120).

We focus now on the aspects of diagnostic competence which were enhanced by the knowledge of, and confident use of, the interviews and growth points.

5.1 A Clearer, Evidence-Based Understanding of Student Thinking in Mathematics and What Students Know and Can Do

Cohen and Ball (1999) argued that "instructional capacity is partly a function of what teachers know students are capable of doing and what they think they are capable of achieving with students" (p. 7). One of the advantages of administering the Early Numeracy Interview at both the beginning and end of the school year is that teachers are provided with exciting evidence of growth in student understanding over time.

Table 1 Percentage success on tasks with small sets

Item	Beginning of first year of school (n = 1,438)	End of first year of school (n = 1,450)
Sort by color	98	100
Count a collection of 4	93	99
Identify one of two groups as "more"	84	99
Make a set, cardinal number 5	85	98
Conserve number	58	88

Table 1 shows the percentage of children on arrival at school (typically five year olds) and at the completion of the school year, respectively, who were able to successfully complete tasks to do with sorting, counting, and conservation (Clarke et al., 2006).

The percentage success on each item increased considerably by the end of the school year. By considering a single class' data and the state data, together with all that they have learned about the individuals in their care, teachers can gain a sense of what typical performance looks like over a year, sometimes in contrast to published curriculum expectations.

5.2 *Realistic Mathematical Expectations of Students*

There was evidence in the ENRP that through extensive use of the interview, teachers developed more realistic expectations of what children knew and could do. Early in the project, teachers made comments in the Surprises and Highlights questionnaire such as, "my greatest surprise was that most children performed significantly better than I anticipated. Their thinking skills and strategies were more sophisticated than I expected" (Clarke et al., 2002, p. 260). In contrast, teachers were surprised with the difficulty that many children appeared to have with tasks relating to abstracting multiplication (Sullivan, Clarke, Cheeseman, & Mulligan, 2001), ordering whole numbers, reading clocks, and identifying the triangles on a page of triangles and non-triangles. An overall change in teachers' diagnostic competence was in their awareness of the considerable range of levels of mathematical understanding in their classes.

This was quantified in the ENRP, when teachers were asked, in the Entry and Exit questionnaires, to indicate whether *none, some, most,* or *all* of their children could do certain tasks. For example, teachers of Preps (five year olds in the first year of school) were asked how many of their children by the end of the year would know that *four hundred and two is written 402 and knows why neither 42 or 4002 is correct.* At the beginning of the project, 61% of the teachers said that none of their children would know that, while at the end of the project, the percentage had dropped to 30% (Clarke et al., 2002). This was a consistent pattern in the data, where teachers, through the use of the interview, were far more likely to indicate that *some* or *most* of their children would know a particular mathematical idea, and far less responded *none* or *all*, evidence that they were far more aware of the diversity of understanding in their classrooms.

5.3 An Understood Framework/Growth Points/Typical Learning Trajectory for Students in a Given Domain

The growth points in the ENRP informed the creation of interview tasks and the recording, scoring, and subsequent data analysis, although the process of development of interview and growth points was very much a cyclical one. In discussions with teachers, we came to describe growth points as key "stepping stones" along the paths to mathematical understanding. They provide a kind of mapping of the conceptual landscape (Fosnot & Dolk, 2002).

To clarify further what is meant by growth points, the six growth points for the ENRP domain of addition and subtraction strategies are shown in Fig. 4.

We do not claim that all growth points are passed by every student along the way. As van den Heuvel-Panhuizen (2001) emphasized, "a teaching-learning trajectory

1. Count-all (two collections)
Counts all to find the total of two collections.
2. Count-on
Counts on from one number to find the total of two collections.
3. Count-back/count-down-to/count-up-from
Given a subtraction situation, chooses appropriately from strategies including count-back, count-down-to and count-up-from.
4. Basic strategies (doubles, commutativity, adding 10, tens facts, other known facts)
Given an addition or subtraction problem, strategies such as doubles, commutativity, adding 10, tens facts, and other known facts are evident.
5. Derived strategies (near doubles, adding 9, build to next ten, fact families, intuitive strategies)
Given an addition or subtraction problem, strategies such as near doubles, adding 9, build to next ten, fact families and intuitive strategies are evident.
6. Extending and applying addition and subtraction using basic, derived and intuitive strategies
Given a range of tasks (including multi-digit numbers), can solve them mentally, using the appropriate strategies and a clear understanding of key concepts.

Fig. 4 ENRP growth points for the domain of addition and subtraction strategies

should not be seen as a strictly linear, step-by-step regime in which each step is necessarily and inexorably followed by the next" (p. 13).

5.4 Revelations About "Quiet Achievers" in the Classroom

In response to a written question on the Highlights and Surprises questionnaire, following their first substantial use of the Early Numeracy Interview, one teacher commented:

> In every class there is that quiet child you feel that you never really 'know'—the one that some days you're never really sure that you have spoken to. To interact one-to-one and really 'talk' to them showed great insight into what kind of child they are and how they think (ENRP teacher March 1999).

A number of teachers noted that the one-to-one interview enabled some "quiet achievers" to emerge, and several noted that many were girls. There appeared to be some children who did not involve themselves publicly in debate and discussion during whole-class or small-group work, but given the individual time with an interested adult, were able to show what they knew and could do.

The experience of the interview meant that many teachers became more sensitive to quiet achievers, and realized that a child not offering much in whole-class discussions did not necessarily mean that they did not have a full understanding of the strategies and concepts being addressed.

5.5 Enhanced Subject Matter Knowledge and Pedagogical Content Knowledge

The evidence from the ENRP and the ACU Rational Number Project indicates that the use of the interviews contributes to enhanced teacher knowledge (Clarke, 2008; Clarke et al., 2002). In the middle years, many teachers acknowledge their lack of a connected understanding of rational number (Lamon, 2007), often using limited subconstructs (sometimes only part-whole) and limited models (such as the ubiquitous "pie"). Many teachers using the rational number interview have reported that their own understanding of rational number (e.g., an awareness of subconstructs of rational number such as measure and division, and the distinction between discrete and continuous models) has been enhanced as they observe the variety of strategies their students draw upon in working on the various tasks. In professional learning settings, we have noticed that a number of middle-school teachers have difficulty in solving the rational number task (ordering ragged decimals) shown in Fig. 2.

Some might presume that teachers' pedagogical content knowledge in the first three years of school would not be an issue. However, many teachers reported that terms such as "counting on," "near doubles," and "dynamic imagery" were unfamiliar to them, prior to their involvement in the ENRP, but came to be shared language. It is interesting to consider whether this is specialized content knowledge or pedagogical content knowledge (see, e.g., Hill, Ball, & Schilling, 2008). As mentioned earlier, it is difficult to categorize exactly the kinds of knowledge which are evident in teachers' practice (Graeber & Tirosh, 2008), but we would argue there is little doubt that both subject matter knowledge and pedagogical content knowledge are enhanced by the use of such interviews.

5.6 An Awareness of Common Strategies Used by Students

In the Teachers' Highlights and Surprises questionnaire data (Clarke, 2008), teachers reported that the training for, and the use of, the interviews enhanced their diagnostic competence by giving them an awareness of strategies in solving problems with which they were not previously familiar. The ACU Rational Number Interview provided examples of this. In the fraction comparison task, students were asked to decide which of two fractions was the larger, for eight pairs, giving reasons for their decisions. These data are discussed in considerable detail in Clarke and Roche (2009). The fraction pairs presented to the student are shown in Fig. 5. Each pair, typed on a card, was placed in front of the student one pair at a time, and the student was asked to point to the larger fraction of the pair, explaining their reasoning. There was no time limit involved.

Researchers report frequently that students use strategies in solving fraction comparison tasks which they are unlikely to have been specifically taught. The use of *residual thinking* (Post & Cramer, 2002) and *benchmarking* (or transitive, Post, Behr, & Lesh, 1986) are likely to be evidence of conceptual understanding and lead to a successful choice.

5.7 An Awareness of Common Difficulties and Misconceptions Demonstrated by Students

As teachers have the opportunity to observe and listen to students' responses, they become aware of common difficulties and misconceptions. For example, many children in the first five years of school (Grades Prep to 4) were unable to give a name to the shape on the left in Fig. 6. It was not expected that they would name it

a) 3/8 7/8	b) 1/2 5/8	c) 4/7 4/5	d) 2/4 4/8
e) 2/4 4/8	f) 3/7 5/8	g) 5/6 7/8	h) 3/4 7/9

Fig. 5 The eight fraction pairs used in the interview (Clarke & Roche, 2009).

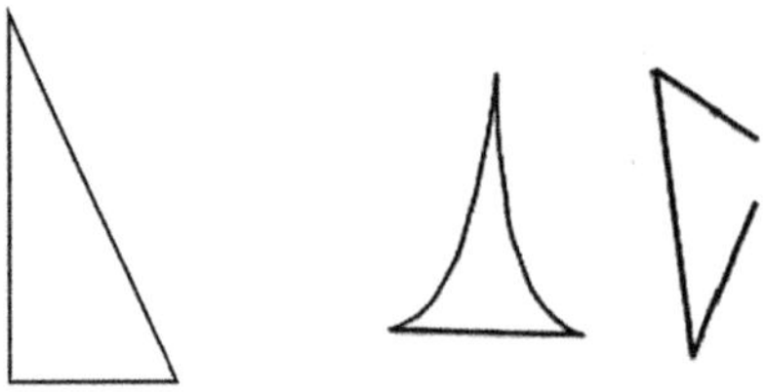

Fig. 6 Triangle and non-triangle shapes in the interview.

"right-angled triangle," but simply "triangle." Because it did not correspond to many students' "prototypical view" (Lehrer & Chazan, 1998) of what a triangle was (i.e., a triangle has a horizontal base and "looks like the roof of a house"—either an isosceles or equilateral triangle), some called it a "half-triangle, because if you put two of them together you get a real triangle." Many students also nominated the two shapes on the right in Fig. 6 as triangles. In fact, in a later task in the interview, 20% of students at the end of Grade 4 were unable to select correctly the triangles from a page of nine shapes (Clarke, 2004).

Following the use of the interview, it was clear from a teaching perspective that it was important to focus on the properties of shapes, and to present students with both examples and non-examples of shapes, as they were coming to terms with definitions.

A common, incorrect strategy in fraction comparison tasks is the use of "gap thinking" (Pearn & Stephens, 2004), often evident in students' responses to task (g) in Fig. 5. Some students claim that 5/6 and 7/8 are equivalent, because they both require one "bit" to make a whole. In this case, the students are focusing on the gap between 5 and 6 and the gap between 7 and 8, but not considering the actual size of the pieces. This gap thinking is really a form of additive rather than proportional thinking, where the student is not considering the size of the denominator and therefore the size of the relevant parts (or the ratio of numerator to denominator), but merely the absolute difference between numerator and denominator.

5.8 *Improved Questioning Techniques, Including the Opportunity to See the Benefits of Increased Wait Time*

Questioning is one of the most important and yet possibly the most challenging aspect for teachers. Shulman (1987) described three critical moments: "when a teacher asks a question he or she knows everyone can answer and no one can! ... when the teacher asks a question he or she is confident that no one can answer—and many do! ... when someone produces an idea or an invention that simply does not fit with the teacher's expectations, and is not immediately discernible as right or wrong" (p. 380).

Researchers studying particularly effective teachers' practice within the ENRP noted that the interview appeared to provide a model for classroom questioning (Clarke & Clarke, 2004). In interviews with the research team, teachers indicated that they found themselves making increasing use of questions of the following kind:

- How did you work that out?
- Can you do it another way?
- How are these two problems the same and how are they different?
- Would that method always work?
- Is there a pattern in your results? (Clarke et al., 2002)

Sleep and Boerst (2012) reported on a program with preservice teachers, designed particularly to enhance teachers' eliciting and interpreting of students' mathematical thinking, in the context of an elementary mathematics methods assignment. Students were required to interview an elementary mathematics student about his or her mathematical thinking, selecting from a mathematical "task pool" focusing on whole numbers and decimals, and writing evidence-based claims about what the student seemed to understand and be able to do. Although, not surprisingly, there were considerable discrepancies between interns' assertions and evidence provided (e.g., assertions being too broad, evidence not specific, a lack of clarity, and evidence contradicting assertion), the authors claimed their work held the potential for advancing tools designed to support practice-based teacher education.

Teachers in the ENRP observed the power of waiting for children's responses during the interview, noting on many occasions the way in which children who initially appeared to have no idea of a solution or strategy, thought long and hard, and then provided a very rich response. Such insights then transferred to classroom situations, with teachers claiming that they were working on allowing greater wait time (Clarke, 2001).

5.9 The Opportunity to Use Tasks from The Interview as Models or Inspirations for Developing Classroom Tasks

The capacity of the teacher to take the information on the record sheet and "map" student performance in relation to the growth points or "big ideas" is a key step in the process of using the interview to inform teaching practice. After conducting the interview, teachers are likely to ask the reasonable question in relation to planning, "so now what?" If they have a clear picture of individual and group performance in particular mathematical domains, they are then in a position, hopefully with the support of colleagues, to plan appropriate classroom experiences for individuals and groups.

Construct a Sum (Fig. 7, Behr, Wachsmuth, & Post, 1985) is an example of where a task used in an assessment interview can be adapted for use as an instructional activity.

The same materials can be used in classrooms with students working in pairs, and inviting them to make the largest sum they can with two fractions, the smallest sum, the sum closest to 3, and so on. In this way, classroom tasks modeled on or inspired by those from the interviews, used together with the kinds of appropriate probing of students' thinking discussed earlier, provided helpful and appropriate

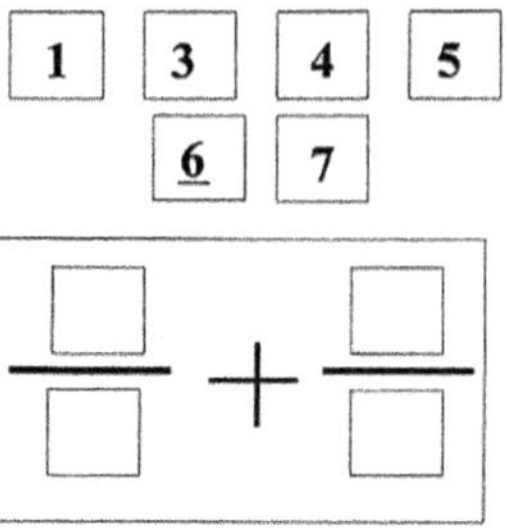

Place the number cards and the empty fraction sum in front of the student.

a) Choose from these numbers to form two fractions that when added together are close to one, but not equal to one. *Record the student's final decision.*

b) Please explain how you know the answer would be close to one.

Fig. 7 Construct a sum task

insights from the interview, and had the potential to lead to the kinds of improved understanding which teachers were seeking.

6 The Use of the Early Numeracy Interview with Children with Down Syndrome

In a recent project that attempted to map the mathematical development of young children with Down syndrome (Faragher & Clarke, 2014), the ENRP interview was adapted and a slightly different approach was taken to its application (Clarke, 2015). Literature indicated that children with Down syndrome interviewed in unfamiliar contexts by people they did not know reduced performance on literacy tasks (Brown & Semple, 1970). Therefore, the researchers interviewed children with Down syndrome in their home or school, in the presence of their parents (or teacher) who watched from behind the child. The adults were invited to comment on the performance of the child, either by taking notes during the interview, or in a discussion following the interview.

In the Down syndrome project, the interview was implemented in a more flexible form than in the ENRP and associated project to ensure maximum opportunities for individual children to show what they knew and could do, rather than as a protocol-driven instrument. Tasks were first asked in the same form of wording as the original instrument but follow-up questioning, instructions, or guidance were provided at the discretion of the interviewer. This allowed the interviewer to follow up on responses from the child, to double back to earlier tasks, to ask a similar task in a different way and to add tasks, such as counting stickers that had been given as rewards during an interview. In order to do this, the interviewer needed to know the purpose behind the interview questions as well as be able to make preliminary judgments about what was being observed in the interview while it was in progress. The interviews were video-taped to allow more detailed analysis.

An example of the more flexible approach to the interview was when one of the questions from the Early Numeracy Interview that focused on location language was asked. The original task asked children to place a small plastic teddy bear in a specified position relative to another teddy. One child was asked to place a green teddy behind the blue teddy bear that was in front of her on the table. She did not do this so the interviewer got out of her seat, moved over to the clear space with the girl and asked her to stand behind her. The child did this successfully, showing some understanding of the concept "behind." This additional task became a feature of future interviews within the Down syndrome project, providing additional information on the mathematical understanding of the children.

7 Conclusion

In this chapter, we have argued that enhanced diagnostic competence gained from the understanding and use of one-to-one interviews has the potential to lead to powerful knowledge in action. We have given examples of how teachers' subject matter knowledge (particularly specialized content knowledge and horizon content knowledge) and pedagogical content knowledge can be enhanced, resulting in the nine benefits discussed in this chapter.

Ginsburg (2009) noted that good teaching involved "understanding the mathematics, the trajectories, the child's mind, the obstacles, and using general principles of instruction to inform the teaching of a child or group of children" (p. 126). We would argue that this chapter provides compelling evidence that the task-based, one-to-one assessment interview can make a major contribution to such understanding, through greatly increasing teachers' diagnostic competence.

References

Behr, M., & Post, T. (1992). Teaching rational number and decimal concepts. In T. Post (Ed.),*Teaching mathematics in grades K-8: Researchbased methods* (2nd ed., pp. 201–248). Boston, MA: Allyn and Bacon.

Behr, M. J., Lesh, R., Post, T. R., & Silver, E. A. (1983). Rational number concepts. In R. Lesh & M. Landau (Eds.), *Acquisition of mathematics concepts and processes* (pp. 91–126). New York, NY: Academic Press.

Behr, M. J., Wachsmuth, I., & Post, T. R. (1985). Construct a sum: A measure of children's understanding of fraction size. *Journal for Research in Mathematics Education, 16*(2), 120–131.

Bobis, J., Clarke, B. A., Clarke, D. M., Gould, P., Thomas, G., Wright, R., & Young-Loveridge, J. (2005). Supporting teachers in the development of young children's mathematical thinking: Three large scale cases. *Mathematics Education Research Journal, 16*(3), 27–57.

Brown, R. I., & Semple, L (1970). Effects of unfamiliarity on the overt verbalisation and perceptual motor behaviour of nursery school children. *British Journal of Educational Psychology, 40*, 291–298.

Clarke, B. A. (2004). A shape is not defined by its shape: Developing young children's geometric understanding. *Journal of Australian Research in Early Childhood Education, 11*(2), 110–127.

Clarke, B. A. (2008). A framework of growth points as a powerful teacher development tool. In D. Tirosh & T. Wood (Eds.), *The international handbook of mathematics teacher education, Tools and processes in mathematics teacher education* (Vol. 2, pp. 235–256). Rotterdam, Netherlands: Sense Publishers.

Clarke, B. A. (2015). Assessing young children's mathematical understanding: Opportunities and expectations at the transition to school. In B. Perry, A. MacDonald, & A. Gervasoni (Eds.), *Mathematics and transition to school* (pp. 31–46). Singapore: Springer.

Clarke, B. A., & Clarke, D. M. (2004). Using questioning to elicit and develop children's mathematical thinking. In G. W. Bright & R. N. Rubenstein (Eds.), *Professional development guidebook for perspectives on the teaching of mathematics* (pp. 5–10). Reston, VA: National Council of Teachers of Mathematics.

Clarke, B. A., Clarke, D. M., & Cheeseman, J. (2006). The mathematical knowledge and understanding young children bring to school. *Mathematics Education Research Journal, 18*(1), 81–107.

Clarke, D., Clarke, B., & Roche, A. (2011). Building teachers' expertise in understanding, assessing and developing children's mathematical thinking: The power of task-based, one-to-one interviews. *ZDM Mathematics Education, 43*(6), 901–913.

Clarke, D. M. (2001). Understanding, assessing, and developing young children's mathematical thinking: Research as a powerful tool for professional growth. In J. Bobis, M. Mitchelmore, & B. Perry (Eds.),*Numeracy and beyond* (Proceedings of the 24th Annual Conference of the Mathematics Education Research Group of Australasia, pp. 9–26). Sydney, Australia: MERGA.

Clarke, D. M., & Roche, A. (2009). Students' fraction comparison strategies as a window into robust understanding and possible pointers for instruction. *Educational Studies in Mathematics, 72*, 127–138.

Clarke, D. M., Cheeseman, J., Gervasoni, A., Gronn, D., Horne, M., McDonough, A., … Rowley, G. (2002). *Early Numeracy Research Project final report.* Melbourne, Australia: Mathematics Teaching and Learning Centre, Australian Catholic University.

Clarke, D. M., Roche, A., Mitchell, A., & Sukenik, M. (2006). Assessing student understanding of fractions using task-based interviews. In J. Novotna, H. Moraova, M. Kratka, & N. Stehlikova (Eds.), *Proceedings of the 30th Conference of the International Group of Psychology of mathematics Education* (Vol. 2, pp. 337–344). Prague, Czech Republic: PME.

Clements, D. H., Swaminathan, S., Hannibal, M. A. Z., & Sarama, J. (1999). Young children's conceptions of space. *Journal for Research in Mathematics Education, 30*(2), 192–212.

Clements, M. A., & Ellerton, N. (1995). Assessing the effectiveness of pencil-and-paper tests for school mathematics. In B. Atweh & S. Flavel (Eds.),*Galtha: MERGA 18* (Proceedings of the 18th Annual Conference of the Mathematics Education Research Group of Australasia, pp. 184–188). Darwin, Australia: MERGA.

Cohen, D. K., & Ball, D. L. (1999). *Instruction, capacity, and improvement.* Philadelphia, PA: Consortium for Policy Research in Education, University of Pennsylvania.

Department of Education & Training. (2001). *Early numeracy interview booklet.* Retrieved from https://www.eduweb.vic.gov.au/edulibrary/public/teachlearn/student/mathscontinuum/onlineinterviewbklet.pdf

Erickson, F. (1986). Qualitative methods in research on teaching. In M. C. Whittrock (Ed.), *Handbook of research on teaching* (pp. 119–161). New York, NY: Macmillan.

Faragher, R., & Clarke, B. A. (Eds.). (2014). *Educating learners with down syndrome: Research, theory and practice with children and adolescents.* London, UK: Routledge

Fosnot, C. T., & Dolk, M. (2002). *Young mathematicians at work: Constructing fractions, decimals and percents.* Portsmouth, NH: Heinemann.

Ginsburg, H. (2009). The challenge of formative assessment in mathematics education: Children's minds, teachers' minds. *Human Development, 52*, 109–128.

Ginsburg, H., Klein, A., & Starkey, P. (1998). The development of children's mathematical thinking: Connecting research with practice. In I. E. Siegel & K. A. Renninger (Eds.), *Handbook*

of child psychology, Child psychology in practice (Vol. 4, 5th ed., pp. 23–26). New York, NY: John Wiley & Sons.
Graeber, A., & Tirosh, D. (2008). Pedagogical content knowledge. In P. Sullivan & T. Wood (Eds.),*Knowledge and beliefs in mathematics teaching and teaching development* (pp. 117–132). Rotterdam, The Netherlands: Sense Publishers.
den Heuvel-Panhuizen, V. (2001). *Children learn mathematics.* Utrecht, The Netherlands: Freudenthal Institute.
Hill, H., Ball, D., & Schilling, S. (2008). Unpacking pedagogical content knowledge: Conceptualising and measuring teachers' topic-specific knowledge of students. *Journal for Research in Mathematics Education, 39*, 372–400.
Horne, M, & Rowley, G. (2001). Measuring growth in early numeracy: Creation of interval scales to monitor development. In M van den Heuvel-Panhuizen (Ed.), *Proceedings of the 25th Conference of the International Group for the Psychology of Mathematics Education* (pp. 3-161–3-168). Utrecht, The Netherlands: Freudenthal Institute.
Lamon, S. (2007). Rational numbers and proportional reasoning. In F. K. Lester (Ed.), *Second handbook on research on mathematics teaching and learning* (pp. 629–668). Reston, VA: National Council of Teachers of Mathematics.
Lehrer, R., & Chazan, D. (1998). *Designing learning environments for developing understanding of geometry and space*. Mahwah, NJ: Lawrence Erlbaum.
McDonough, A., Clarke, B. A., & Clarke, D. M. (2002). Understanding assessing and developing young children's mathematical thinking: The power of the one-to-one interview for preservice teachers in providing insights into appropriate pedagogical practices. *International Journal of Educational Research, 37*, 107–112.
Miles, M. B., & Huberman, A. M. (1994). *Qualitative data analysis: An expanded sourcebook.* Beverley Hills, CA: Sage.
Pearn, C., & Stephens, M. (2004). Why you have to probe to discover what year 8 students really think about fractions. In I. Putt, R. Faragher, & M. McLean (Eds.), *Mathematics education for the third millennium: Towards 2010 (Proceedings of the 27th Annual Conference of the Mathematics Education Research Group of Australasia)* (pp. 430–437). Townsville, Austraia: MERGA.
Post, T., Behr, M. J., & Lesh, R. (1986). Research-based observations about children's learning of rational number concepts. *Focus on Learning Problems in Mathematics, 8*(1), 39–48.
Post, T., & Cramer, K. A. (2002). Children's strategies in ordering rational numbers. In D. L. Chambers (Ed.), *Putting research into practice in the elementary grades* (pp. 141–144). Reston, VA: National Council of the Teachers of Mathematics.
Roche, A. (2005). Longer is larger—Or is it? *Australian Primary Mathematics Classroom, 10*(3), 11–16.
Roche, A., & Clarke, D. M. (2004). When does successful comparison of decimals reflect conceptual understanding? In I. Putt, R. Farragher, & M. McLean (Eds.), *Mathematics education for the third millennium: Towards 2010* (Proceedings of the 27th annual conference of the Mathematics Education Research Group of Australasia, pp. 486–493). Townsville, Australia: MERGA.
Shulman, L. (1987). The wisdom of practice: Managing complexity in medicine and teaching. In D. C. Berliner & B. V. Rosenshine (Eds.), *Talks to teachers: A festschrift for N. L. Gage* (pp. 369–387). New York, NY: Random House.
Sleep, L., & Boerst, T. A. (2012). Preparing beginning teachers to elicit and interpret students' mathematical thinking. *Teaching and Teacher Education, 28*, 1038–1048.
Sowder, J. (2007). The mathematics education and development of teachers. In F. K. Lester (Ed.), *Second handbook of research on mathematics teaching and learning* (pp. 157–223). Charlotte, NC: Information Age Publishing & National Council of Teachers of Mathematics.
Stake, R. (1995). *The art of case study research*. Thousand Oaks, CA: Sage.
Steinle, V., & Stacey, K. (2003). Grade-related trends in the prevalence and persistence of decimal misconceptions. In N. Pateman, B. Dougherty, & J. Zilliox (Eds.), *Proceedings of the 2003*

Joint Meeting of PME and PMENA (Vol. 4, pp. 259–266). Honolulu, Hawaii: International Group for the Psychology of Mathematics Education.

Sullivan, P., Clarke, D. M., Cheeseman, J., & Mulligan, J. (2001). Moving beyond physical models in learning multiplicative reasoning. In M. van den Heuvel-Panhuizen (Ed.). *Proceedings of the 25th annual conference of the International Group for the Psychology of Mathematics Education* (Vol. 4, pp. 233–240). Utrecht, The Netherlands: Freundenthal Institute.

Swan, P. A. J. (2002). *The computation choices made by students in Years 5 to 7.* Unpublished doctoral dissertation, Edith Cowan University, Western Australia.

Webb, N., & Romberg, T. A. (1992). Implications of the NCTM Standards for mathematics assessment. In T. A. Romberg (Ed.), *Mathematics assessment and evaluation: Imperatives for mathematics education* (pp. 37–60). Albany, NY: State University of New York Press.

Diagnosing Learning Goals: An Often-Overlooked Teaching Competency

James Hiebert, Anne K. Morris, and Sandy M. Spitzer

1 Introduction

To make sense of the phrase we used in the title, "diagnosing learning goals," we need to expose two common misunderstandings. The first misunderstanding is that learning goals are only statements of value, that is, only statements of what the community most wants its students to learn. The second misunderstanding is that diagnosis applies only to assessing students' learning. The purpose of challenging these conventional conceptions of "learning goals" and "diagnosis" is to enable a broader definition of *diagnosing learning goals*, a definition we believe describes a teaching competency often overlooked by both teachers and teacher educators. In this chapter, we describe what diagnosing learning goals could mean by defining the phrase in a somewhat unconventional way. Then, we illustrate how the phrase identifies a competency we have found critical in learning to teach well, and offer a comment about whether teachers and teacher educators can learn to diagnose learning goals.

1.1 Learning Goals Are More than Statements of Value

Learning goals, as traditionally defined, are statements about what academic achievements are most valued by society (Hiebert, 1999). These statements often come in the form of standards (e.g., National Council of Teachers of Mathematics

J. Hiebert (✉) • A.K. Morris
University of Delaware, Newark, DE, USA
e-mail: hiebert@udel.edu; abmorris@udel.edu

S.M. Spitzer
Towson University, Towson, MD, USA
e-mail: sspitzer@towson.edu

T. Leuders et al. (eds.), *Diagnostic Competence of Mathematics Teachers*, Mathematics Teacher Education 11, DOI 10.1007/978-3-319-66327-2_10

(NCTM), 1989, 2000; National Governors Association Center for Best Practices & Council of Chief State School Officers. (CCSSM, 2010). Standards usually are phrased at a general level—they describe what students are to learn by the end of the year, or by the end of studying a particular topic. They rarely are written at a specific enough level to guide instruction day to day, at the lesson level.

Learning goals can be defined at a more specific level—at a level that can guide teachers' daily instruction. We believe that learning goals at this level are best conceived as hypotheses about what specific ideas students must learn to achieve the broader, value-based goals. For example, "learning the meaning of operations with fractions" often emerges from value judgments, whereas "learning that $a/b \div c/d$ means how many copies of c/d can be removed from a/b" is a hypothesis about what students must learn to understand the meaning of dividing fractions.

In this chapter, we are interested in these more specific learning goals, those that suggest what kind of instruction to design. We argue that these goals, or hypotheses, can be treated as empirical objects. They can be created, tested, and refined based on evidence collected during instruction. Evidence of students' learning can be used to answer questions like, "Did achieving this particular goal move students toward achieving the larger goal? Is something missing that must be added as a learning goal? Did students show some unanticipated misconception that must be confronted as a separate learning goal?" We claim that treating learning goals as hypotheses that can be tested empirically adds a critical dimension to the meaning of "learning goal" and defines an important competency for teaching, and for improving teaching.

1.2 Diagnosing Is More than Assessing Students' Learning

The title of this book suggests that readers will encounter various ways in which "diagnostic competence" can be defined and that unpacking this idea will reveal important competencies for teaching mathematics. Traditionally, "diagnosis" applied to mathematics teaching and learning has been treated as assessing learning or achievement. Many of the chapters in this book deal with the various ways in which this meaning of diagnosis can be understood and practiced. Although diagnosing students' learning is a complex competency and warrants lots of unpacking, this chapter is about something different. It is about diagnosing learning goals.

Diagnosing learning goals, as we will define it, makes sense only if learning goals are treated as hypotheses (as described in the previous section). But, one more clarification will help the reader better understand what we have in mind. Many dictionaries contain, as the primary definition of diagnosis, the identification of a medical problem. These dictionaries often contain, as a secondary definition, something like "Investigation or analysis of the cause or nature of a condition, situation, or problem" (Merriam-Webster Dictionary, 2015). If we use this second definition of diagnosis, then "diagnosing learning goals" becomes "analyzing learning goals." The analysis of learning goals is exactly what we have in mind. Analyzing learning goals, from a mathematical point of view, requires first breaking them into their constituent parts or subgoals. By this, we mean specifying, as precisely as possible, the mathematical ideas embedded in the goal. Analysis continues by hypothesizing which subgoals students must understand in order to achieve the broader (often value-determined) learning goal, testing whether the hypotheses are confirmed, and revising the learning goal or its subgoals. Such revisions might add a subgoal that was omitted or rephrase a goal that guided instruction in a non-productive direction. Through this analysis and refinement of learning goals, teachers and teacher educators can improve their instruction and its alignment with their broader goals for students' learning.

2 Setting the Stage for Treating Learning Goals as Empirical Objects

For the past 15 years, mathematics faculty and doctoral students in the School of Education, University of Delaware (UD), have been working systematically to improve the mathematics portion of the K-6 teacher preparation program. The program graduates about 130 students every year. The mathematics portion of the program consists of three mathematics content courses and one mathematics methods course. Students in the program take the three-semester-long sequence of mathematics courses as freshmen and sophomores, and the methods course as juniors or seniors. Several sections of each course are offered each semester; faculty and doctoral students serve as instructors. Each semester, the team of two to four instructors for each course meets weekly to study and improve small parts of the course that remain problematic (a single lesson or instructional activity). All instructors teach from the same detailed lesson plans, so they all teach toward the same learning goals using similar methods. This enables meaningful shared analyses of the effects of specific instruction on our students' achievement of the lesson-level learning goals (see Hiebert & Morris, 2009, for a more complete description of these activities at the University of Delaware).

We see the process we are using as an example of "Improvement Science" (Kenney, 2008; Langley et al., 1996; Morris & Hiebert, 2009) applied to improving teaching. In simple terms, our application of Improvement Science involves repeated cycles of formulating hypotheses about what pre-service teachers (PSTs) need to know to achieve the broader learning goals, testing these hypotheses, making changes to the lesson or refining the hypotheses about the critical subgoals, testing again, and so on. Some readers will, appropriately, recognize features of "lesson study" in this process (Arani, Keisuke, & Lassegard, 2010; Fernandez & Yoshida, 2004; Lewis, 2002). We have formalized the version of the process we use, and proposed it as a continuous improvement model for any substantive educational goal, by identifying four critical features (Morris & Hiebert, 2011):

(1) First, the process is intended to create shared "instructional products." In our case, these products are lesson plans. These products should contain the growing

knowledge about how to help participants achieve the learning goals. Because they are written, concrete products, the knowledge they contain can be passed along to future instructors; knowledge can build and accumulate rather than being lost when experienced instructors leave.

(2) The second feature of the model is setting clear, shared, stable learning goals. The goals must be clear and explicit so their achievement can be measured; they must be shared so instructors can conduct meaningful tests of the learning goals across sections of the courses; and they must be stable so knowledge can accumulate and instruction can improve over time. The need for stability means these goals are the larger, value-determined goals; specific, sub-concept goals will change as more is learned about how to help students achieve the larger goals.

(3) The third feature of our model is using small tests of small changes to test the hypotheses. Collecting just enough data to tell whether a hypothesized learning subgoal guides instruction in the right direction allows improvements to continue and momentum to be sustained.

(4) The final feature in our model is the solicitation of ideas and suggestions from everyone involved in the process. To create productive innovations, any improvement system needs a steady flow of new ideas.

Diagnosing learning goals is the focus of this chapter, but it is useful to see this competency as one element in the context of a continuous improvement process. Similarly, it is helpful to see that small tests of small changes is a mechanism that can drive identifying, testing, and refining subgoals that make up the learning goals.

3 Empirically Testing and Improving Learning Goals

3.1 The Process at Work in Our Teacher Preparation Program

In our work at the University of Delaware, instructional products are the lesson plans for each session of each course. Each lesson plan begins with learning goals for that lesson, stated as precisely as possible. The learning goals were initially generated by analyzing the mathematical skills and understandings (e.g., learning that $a/b \div c/d$ means how many copies of c/d can be removed from a/b) we believed were required for our students to achieve the larger learning goals (e.g., understanding why the procedures for dividing fractions work like they do). Our analysis benefitted from reviewing the learning and teaching literature on how students tend to achieve these goals.

Over time, the learning goals have been revised and refined to create more accurate descriptions of all the component parts or subgoals we believe our students, who are pre-service teachers (PSTs), need to know. The goals have also been rephrased more precisely to facilitate the development of more accurate measures of whether PSTs are achieving the goals. Today's learning subgoals are our best current hypotheses about what PSTs must learn to achieve the main lesson-level learning goal(s) and the broader goal(s) for which the lesson goal(s) and subgoals are components. More precise and targeted learning goals are nested within increasingly broad learning goals. Each subgoal serves as a small piece of the foundation of students' understanding of big ideas about mathematics.

As an example, consider the goal of understanding the algorithm for multi-digit multiplication. This is a learning goal of one of our lessons for PSTs. Over time, we have analyzed, or diagnosed, this learning goal into the following subgoals: (1) understanding the meaning of multiplying a times b as creating a copies of b, or a groups of b; (2) understanding how place value affects the number of copies, or size of a copy, by powers of ten (e.g., understanding that when multiplying 53 times 24, "5" copies of 24 means 50 copies of 24); and, (3) understanding the way in which the distributive property creates partial products which can be summed to find the total. Other subgoals could be identified, but we have found these subgoals guide the design of instructional activities which effectively support PSTs' achievement of the larger, lesson goal.

What does this work of diagnosing learning goals entail? To a large extent, diagnosing learning goals requires decomposing broader goals into increasingly precise learning goals that can be used to guide the development of lessons, even instructional activities within lessons. The adequacy of the diagnosis is determined by whether the PSTs achieve the lesson goal and, if not, by identifying what understanding still is needed to do so. In other terms, the diagnostic competence we are describing consists of identifying the ways in which students' mathematical thinking could be improved and then unpacking the relevant mathematics learning goal into component parts that reveal the mathematics that could be further addressed in better-designed instructional activities.

How have we supported the development of the competency of diagnosing learning goals in our teacher preparation program? Before answering this question, we need to remind the reader that, in this section, we are describing the development of competency among instructors of PSTs. So, we, as faculty at the University of Delaware, along with our doctoral students, are the teachers, and PSTs are the students. In the latter sections of the chapter, we will describe this same competency among PSTs and offer some suggestions about what it will take to help pre-service and novice teachers develop the same competency.

Among instructors in our teacher preparation program, developing the ability to analyze learning goals is embedded within the continuing cycles of improving lessons for PSTs. It is not isolated as a separate skill to be taught to our course instructors apart from improving teaching. Each cycle of improvement begins with diagnosing the primary learning goal of a lesson into a set of hypothesized subgoals that state what PSTs must understand and be able to do to achieve the larger goal. At this point in our process, lessons already contain these subgoals, so most cycles of improvement involve looking at the performance of PSTs the previous semester and asking whether changes are needed to the statements of the subgoals and/or whether better-aligned instructional activities should be designed.

The cycle of improvement continues by making predictions about the effects of the (possibly) revised activities on PSTs' achievement of the (possibly) revised subgoals. The predictions are then tested by collecting data on PSTs' learning. Then, those data are interpreted to confirm the appropriateness of the activities and/or the subgoals, reveal any missing or incomplete subgoals, and revise the activities and/or subgoals as needed.

To make this process concrete, we present three examples of this process at work in our teacher preparation program. The examples serve two purposes: (1) illustrate the way in which instructors of our pre-service courses develop competence in analyzing learning goals (and the contexts that support this analysis); and, (2) demonstrate how research can be conducted to uncover the mechanisms at work in developing this competence.

Example 1

Morris and Hiebert (2015) report on two groups of instructors, one during the fall semester and a second, different group during the following spring semester. The groups of instructors included faculty and doctoral students in mathematics education. Both groups taught the same course—the first of the three mathematics content courses for PSTs. Morris conducted a study that followed the two groups during their weekly instructor meetings. The purpose was to uncover the mechanisms that motivated instructors to analyze learning goals for the lessons and create additional subgoals when PSTs failed to achieve the lesson-learning goals. To preview a complex set of results, instructors spontaneously asked what subgoals might be missing from the primary learning goal only when they looked back at the performance of the PSTs during a previous semester in the same course and noticed a pattern of poorer-than-expected performance across several lessons.

During the first stage of the study, the fall instructors collected small amounts of data to assess the effectiveness of particular lesson activities. To do this, they used a few multiple choice items that contained seductive distractors to check whether the PSTs had achieved the learning subgoals for a previous lesson. The items were administered at the beginning of the following lesson. To save time, all PSTs had electronic responders and could push the button on their responder to give their choice for each item. Choices for the class of about 30 PSTs were tallied electronically and projected on a screen in the form of a bar graph to show the performance of the class, as a whole. We used these items, sometimes called "clicker items," to test whether PSTs had achieved particular subgoals from the previous lesson. These data were saved and can be used by instructors to improve the lessons over time.

Morris learned that the fall instructors had trouble using data showing poor performance on a previous lesson to analyze and improve the lesson because it is difficult to look back and create testable hypotheses about what caused the poor performance—there are too many possibilities. However, when the spring instructors looked across several lessons in this unit, they noticed a pattern in the fall data: PSTs were not correctly interpreting some of the diagrams used to show the meaning of arithmetic operations with whole numbers and decimals. For example, for the

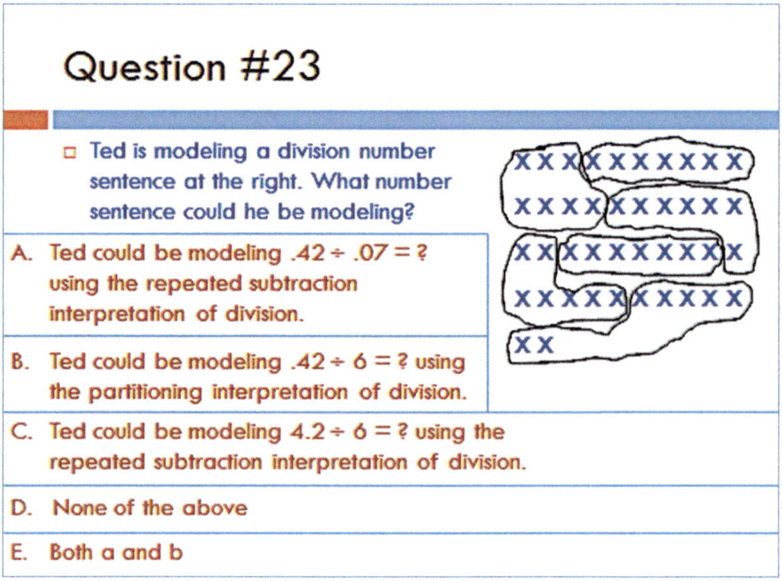

Fig. 1 Item example

clicker item in Fig. 1, the percentage of correct responses ranged from 14% to 53% in the four sections of the course. The most common response was A; the correct response is E.

The spring instructors hypothesized that PSTs were likely missing specific competencies needed to understand and work flexibly with diagrams. In other words, the spring instructors analyzed the learning goals and decided specific subgoals were missing—learning how to interpret these kinds of diagrams. Because they had not yet taught these lessons, the instructors could add these subgoals, design instructional activities to address the subgoals, assess whether the PSTs achieved the subgoals after engaging in these activities, compare performance to the fall semester, and revise their subgoals and/or the instructional activities.

Through several cycles of this process (studying instructional activities as different lessons introduced diagrams for different operations), the spring instructors refined their statement of the subgoals required to interpret diagrams. By the time they taught division, the instructors hypothesized that interpreting diagrams appropriately requires achieving some subgoals that cut across all arithmetic operations (e.g., understanding some number sentences can be modeled in multiple ways, understanding diagrams are ambiguous if the units of measure are not made explicit) and achieving some subgoals unique to a particular operation (e.g., division has two different meanings—partitioning and repeated subtraction). Diagnosis of the learning goal led these instructors to directly target the subgoals with which the PSTs struggled and gave these instructors insights into how the structure of the mathematics influences learning.

We believe what encouraged the spring instructors to use data to re-analyze learning goals was the luxury of looking back at past performance, diagnosing which learning subgoals were missing, inserting these subgoals with aligned instructional activities, and predicting the performance of their students. *Predicting* performance based on revisions seemed to be the key to motivating instructors to engage in a series of improvement cycles in which diagnosing learning goals played a major role.

After observing the spring instructors' work to improve their teaching of diagraming arithmetic operations, Morris and Hiebert (2015) conclude:

"Stepping back, we believe a key lesson to draw from the cycles of hypothesis testing is that decomposition of an initial, often vaguely defined learning goal into its component parts is a critical outcome of hypothesis formulation and refinement during sequential cycles of testing. Unless teachers can unpack often generally stated learning goals into more precise component parts, they cannot design appropriate instruction, they cannot focus their improvement efforts in productive ways, and they cannot measure precisely enough students' performance to know if their changes are actual improvements" (p. 137).

Example 2

Berk and Hiebert (2009) also describe a cycle of improvement within a mathematics course for PSTs. In this case, the course was the second of the three content courses in the preparation program at the University of Delaware. The process of building competence in diagnosing learning goals among the teacher educators worked somewhat differently than in Example 1. But, at the core, the process of diagnosing learning goals still involved unpacking the mathematics in the lesson goal, analyzing and revising the learning goal(s) for the lesson based on this analysis, and observing the PSTs' responses to the revised instructional activities designed to align more closely to the new subgoals. The report of Berk and Hiebert (2009) spanned three semesters.

The first group of instructors taught a lesson in which one of the learning goals was stated as "Prospective teachers will understand how to represent subtraction of fractions with a story problem" (ibid., p. 344). Observations by the instructors, shared during the weekly meeting, all pointed to a failure of many PSTs to achieve this goal. Evidence from course exams confirmed the instructor observations: the learning goal on writing story problems was not achieved by most PSTs. Notes were written in the plan for this lesson suggesting to the following semester's instructors that they analyze this difficulty further, develop hypotheses about the reasons for PSTs' failures, and test their hypotheses with carefully targeted instructional activities.

Based on the notes left by the first group, the group of instructors for the same course the following semester hypothesized that the difficulty emerged from PSTs' failure to understand the role of the referent (i.e., what counts as a unit). They recognized the learning goal did not explicitly mention understanding the referent. Consequently, they inserted a new subgoal so the lesson learning goal now read "Prospective teachers will understand how to represent subtraction of fractions with

Table 1 Alignment of subtraction of fraction number sentences and story problems

Uses the same referent	Does not use the same referent
Kathy has ½ pound of chocolate. She eats ¼ pound of the chocolate. How much chocolate is left?	Kathy has ½ pound of chocolate. She eats ¼ of the chocolate. How much chocolate is left?
$\frac{1}{2}-\frac{1}{4}=?$	$\frac{1}{2}-\left(\frac{1}{4}\times\frac{1}{2}\right)=?$

a story problem. This involves understanding the need to employ the same referent for each fraction as well as being able to distinguish story problems in which the referent is the same from those in which the referent is different" (Berk & Hiebert, 2009, p. 347). Working from the elaborated learning goal, these instructors revised the instructional activity to place a sharper focus on the importance of attending to the referent. An analysis of PSTs' written responses to the instructional task showed a majority of them now attended to the referent in writing story problems. However, an analysis of the videotapes taken of this lesson showed that some PSTs still struggled with this concept. Their written responses confirmed their difficulties.

The third semester's instructors, armed with the information written in the lesson plan by the second group, revised the learning goal still further. They added, as a separate subgoal, the ability to align story problems and number sentences using the fact that some employ the same referent and some do not (see Table 1 for an example). This group of instructors altered the instructional activities again to focus attention on the different types of stories and the number sentences for each. They created the instructional activity so that many PSTs initially made the anticipated mistake. The activity then asked PSTs to consider pairs of stories like those in Table 1 and write number sentences for each. This contrast, suggested as instructionally beneficial by the statement of the revised learning subgoal, seemed to effectively help almost all PSTs recognize the importance of the referent when writing story problems for fraction subtraction problems.

As in the first example, instructors (faculty and doctoral students) of the teacher preparation courses developed their competency for diagnosing learning goals as a natural part of the cycle of improvement applied to improving the lesson plans for these courses. The competency involved analyzing the mathematics contained in the statement of the learning goal and observing PSTs' responses to new or revised instructional activities designed to help PSTs achieve the new subgoals.

Example 3

A third example from our efforts to improve our own competency in diagnosing learning goals comes from the mathematics methods course for our PSTs (Jansen, Bartell, & Berk, 2009). In this example, the learning goal we needed to unpack was pedagogical rather than mathematical, but the same procedures apply. A key course-level learning goal was "Prospective teachers will assess students' thinking to evaluate the effects of their instruction" (p. 531). In this example, we were diagnosing the adequacy of a learning goal for PSTs; however, the goal, itself, focused on assisting PSTs to develop a similar diagnostic competency.

We knew, from past research, that PSTs often evaluate the effectiveness of their instruction based on whether they demonstrate particular behaviors rather than whether their students change their thinking in intended ways. Jansen and Spitzer (2009) had conducted a study to determine, in more detail, how our prospective teachers evaluated the effects of their own instruction given the instructional activities of the course designed with the initial, rather general goal quoted above. Results showed that PSTs chose to write more carefully about their students' thinking when they attended to individual students rather than the class as a whole, and when they described students' responses during instruction as more than just right or wrong. Based on these findings, the learning goal for the methods course was unpacked and new subgoals were inserted. The elaborated learning goal now read "When assessing students' thinking to evaluate the effects of their instruction, prospective teachers will characterize students' thinking by describing students' mathematical thinking and differentiating between students" (Jansen et al., 2009, p. 531). In other words, we were learning, by conducting research on the effectiveness of our methods course, how we could create more precise subgoals that would help our PSTs begin developing the same competency we were developing—diagnosing learning goals.

This third example shows that learning goals can be refined through single studies as well as through repeated cycles of small tests of small changes across multiple semesters. The example also shows that pedagogical, as well as mathematical, goals can be the target of diagnosis and can be part of cycles of improvement of teaching. In all cases, we believe the diagnosis of learning goals, and the subsequent changes in instruction, never reach a finished state. Learning goals, and instruction, can always be further diagnosed, refined, and improved.

3.2 Summary

Our intent in describing the continuous refinement and elaboration of learning goals in the lesson plans for our teacher preparation courses is to demonstrate the way in which the diagnosis of learning goals is an integral part of efforts to improve teaching. In the examples we presented, teaching is the instruction of PSTs by teacher educators, but we see no relevant difference in this setting and in the improvement of classroom teaching. In our view, improving teaching must treat the diagnosis of lesson-level learning goals as a process of formulating hypotheses about subgoals that state what students need to know and do to achieve the larger learning goals (often driven by value judgments), and testing these hypotheses against student learning data. In our view, developing the competence to diagnose learning goals is an integral part of the repeated cycles of improvement targeted toward improving classroom teaching. It is not a competency that should be isolated and trained as a separate skill. This is true whether the teachers are instructors of teacher education courses or classroom teachers.

We mentioned earlier that the process of improving teaching we have described shares many features with lesson study. Readers might also have noticed similarities of this process with "design research" (Cobb, Confrey, diSessa, Lehrer, & Schauble, 2003; Gravemeijer & van Eerde, 2009). Both the cycles of improvement we describe and the process of design research engage in recurring efforts to improve instruction to help students achieve particular learning goals. One difference we see between the two approaches is that design research concludes with a refined local theory of best instruction whereas the cycles of improvement we describe have no end. There is no point at which instruction is good enough; improvements can always be made as more precise and well-defined learning goals are identified.

We conclude this section by noting that, based on our experience as instructors of teacher education courses, two kinds of knowledge or skills are especially important for developing competence to diagnose learning goals. One is the knowledge needed to unpack learning goals into component parts. In most of our work, the learning goals have been mathematical and the knowledge needed is part of mathematical-knowledge-for-teaching described by Ball, Thames, and Phelps (2008). A second skill is observing students' thinking in order to identify the nature of the inadequacies or incompleteness that could be enriched by addressing a particular piece of mathematics, a particular mathematics learning (sub)goal. Pinpointing students' incomplete mathematical thinking allows teachers to hypothesize instructional activities that might specifically address these inadequacies.

4 Can Preservice Teachers Diagnose Learning Goals?

Not surprisingly, as teacher educators who have been involved with the continuous diagnosis and refinement of learning goals to improve our preparation courses for elementary mathematics teachers, we began asking ourselves whether our PSTs could diagnose learning goals. Example 3, presented earlier, shows an early effort to help PSTs begin to acquire the foundational competencies needed (e.g., attending to student thinking). We now ask whether PSTs possess the mathematical knowledge needed to unpack the key mathematics learning goals of the elementary school curriculum? Based on the research we describe in the following paragraphs, the answer, in brief, is that PSTs can acquire the mathematics competence necessary to analyze learning goals into constituent parts, but they rarely do so when asked to analyze or improve teaching (Morris, Hiebert, & Spitzer, 2009).

Morris et al. (2009) asked 30 PSTs to complete four tasks. All PSTs had completed the courses that covered the mathematics material needed for these tasks. To set up the first task, PSTs were presented with a learning goal, "Students will understand how to add fractions and will understand the concepts underlying the operation" (p. 497). They were then asked to write an ideal student response to four different problems, each using the expression 1/4 + 3/8 but each asking students to use a different solution method (e.g., “Solve 1/4 + 3/8 by drawing a diagram on paper”). PSTs were told “Imagine this is the only problem Sue will solve for you; that is, this is the only evidence you will use to judge whether Sue understands the concepts underlying the addition of fractions. Use the exact wording that you want Sue to use while she solves the task” (p. 497). PSTs’ responses were scored based on the number of appropriate subgoals they included in their response, that is, in terms of how precisely they were able to diagnose the learning goal and reflect that diagnosis in a student response.

The three remaining tasks in this study, each with its own learning goal, asked PSTs to, respectively, evaluate a student’s incorrect response given during a classroom discussion (identify as many subgoals as you can that you think the student does not understand about the learning goal), evaluate a student’s correct written responses to several problems (use the subgoals or component parts of the learning goal to evaluate what the student clearly understands and might not understand), and analyze a classroom lesson (evaluate the effectiveness of a lesson by reading a transcript, change part of the lesson you think is not effective, and justify your changes based on improving the students’ opportunities to achieve the learning goal). Across these three tasks, PSTs were prompted to analyze the learning goals into their subgoals and then use this analysis to evaluate the students’ understanding (or lack of understanding) of those subgoals, as well as evaluate the effectiveness of instruction in addressing them.

The overall finding of Morris et al. (2009) was that our PSTs demonstrated an ability to identify appropriate subgoals for the four different learning goals contained in the four different tasks (one goal per task) *when* the context was supportive. However, they did not *use* this knowledge to analyze the learning goal and then evaluate the quality of a lesson, the effectiveness of an instructional activity, or students’ thinking. By “supportive context,” we mean a context that makes the subgoal visible while analyzing the mathematical problem. This occurs when, for example, the process of solving a problem makes it clear that a subgoal must be achieved to complete the solution.

5 Implications for Teacher Education

The findings from Morris et al. (2009) indicate that, although PSTs possessed the mathematical knowledge needed to analyze a learning goal into its constituent parts, or its subgoals, they tended not to use this knowledge to assess students’ thinking or analyze the quality of instructional activities. This means they did not spontaneously connect students’ failure to achieve a learning goal with an incompletely developed or missing subgoal. Consequently, they did not formulate and test hypotheses about possible missing mathematical understandings. If PSTs do not use this strategy for improving teaching, then, we believe, as proposed in our model for teaching improvement, they lack *the* core strategy for improvement.

In our experience, diagnosing learning goals, as we have described the process in this chapter, is a difficult task, even for practicing teachers. It is not a skill that comes easily or naturally. This is true even though it is an essential part of engaging in cycles of improvement, something teachers do quite naturally, but often haphazardly and unsystematically. To include diagnosing learning goals as a conscious, deliberate part of continuous improvement, PSTs (and practicing teachers) likely need instruction, and deliberate practice (Ericsson, 2006) in this activity. It seems to be one of those teaching competencies that is not immediately apparent or intuitive (Murray, 1996). It is a competency that is unique to the teaching profession; it is not learned in common mathematics courses. If it is learned at all, it must be included as an explicit learning goal in teacher preparation courses or professional development programs.

The good news is that the mathematical analysis needed to diagnose learning goals—analyze and unpack them into their constituent mathematical parts—is a skill that should be teachable in a teacher preparation program. Many teaching competencies lie outside the purview of teacher preparation programs because they require continuing and immediate classroom contexts. For example, observing students' thinking requires time in the classroom, either as part of pre-service clinical experiences or when graduates begin teaching. But the mathematical analysis involved in diagnosing learning goals is primarily based on deep knowledge of the mathematics referenced in the learning goal. This kind of knowledge can be acquired through well-developed courses in a teacher preparation program. In addition, PSTs can be taught to use this knowledge to improve their competencies to diagnose learning goals (Meikle, 2014).

What remains for our teacher education group is to set the diagnosis of learning goals as a more explicit and specific learning goal for our PSTs. Given our experience as teacher educators, this learning goal then must be subject to the same empirical study as other learning goals. In other words, we conclude this chapter with a message for ourselves: apply to this learning goal the knowledge and skills we have acquired to continuously improving the diagnosis of learning goals. Use this process, along with the design and testing of instructional activities suggested by the diagnosis, to help our PSTs achieve *this* goal more effectively.

Acknowledgement The collection and analysis of data presented in this chapter were supported by the National Science Foundation (Grant # 0084329). The opinions expressed in the chapter are those of the authors and not necessarily those of the Foundation.

References

Arani, M. R. S., Keisuke, F., & Lassegard, J. P. (2010). "Lesson study" as professional culture in Japanese schools: An historical perspective on elementary classroom practices. *Japan Review*, *22*, 171–200.

Ball, D. L., Thames, M. H., & Phelps, G. (2008). Content knowledge for teaching: What makes it special? *Journal of Teacher Education*, *59*(5), 389–407.

Berk, D., & Hiebert, J. (2009). Improving the mathematics preparation of elementary teachers, one lesson at a time. *Teachers and Teaching: Theory and Practice*, *15*(3), 337–356.

Cobb, P., Confrey, J., diSessa, A., Lehrer, R., & Schauble, L. (2003). Design experiments in educational research. *Educational Researcher*, *32*(1), 9–13.

Ericsson, K. A. (2006). The influence of experience and deliberate practice on the development of superior expert performance. In K. A. Ericsson (Ed.), *The Cambridge handbook of expertise and expert performance* (pp. 685–705). New York, NY: Cambridge University Press.

Fernandez, C., & Yoshida, M. (2004). *Lesson study: A Japanese approach to improving mathematics teaching and learning*. New York, NY: Routledge.

Gravemeijer, K., & van Eerde, D. (2009). Design research as a means for building a knowledge base for teachers and teaching in mathematics education. *The Elementary School Journal*, *109*, 510–524.

Hiebert, J. (1999). Relationships between research and the NCTM Standards. *Journal for Research in Mathematics Education*, *30*, 3–19.

Hiebert, J., & Morris, A. K. (2009). Building a knowledge base for teacher education: An experience in K-8 mathematics teacher preparation. *The Elementary School Journal*, *109*, 475–490.

Jansen, A., Bartell, T., & Berk, D. (2009). The role of learning goals in building a knowledge base for elementary mathematics teacher education. *The Elementary School Journal*, *109*(5), 525–536.

Jansen, A., & Spitzer, S. M. (2009). Prospective middle school mathematics teachers' reflective thinking skills: Descriptions of their students' thinking and interpretations of their teaching. *Journal of Mathematics Teacher Education*, *12*(2), 133–151.

Kenney, C. (2008). *The best practice: How the new quality movement is transforming medicine*. New York, NY: Public Affairs.

Langley, G. J., Nolan, K. M., Nolan, T. W., Norman, C. L., & Provost, L. P. (1996). *The improvement guide: A practical approach to enhancing organizational performance*. San Francisco, CA: Jossey-Bass.

Lewis, C. C. (2002). *Lesson study: A handbook of teacher-led instructional change*. Philadelphia, PA: Research for Better Schools, Inc.

Meikle, E. (2014). Preservice teachers' competencies to select and sequence students' solution strategies for productive whole-class discussions. *Mathematics Teacher Educator*, *3*(1), 27–57.

Merriam-Webster Dictionary. (2015). http://www.merriam-webster.com/

Morris, A. K., & Hiebert, J. (2009). Introduction: Building knowledge bases and improving systems of practice. *The Elementary School Journal*, *109*, 429–441.

Morris, A. K., & Hiebert, J. (2011). Creating shared instructional products: An alternative approach to improving teaching. *Educational Researcher*, *40*(1), 5–14.

Morris, A. K., & Hiebert, J. (2015). Openness and measurement: Two principles for improving educational practice and shared instructional products. *Mathematics Teacher Educator*, *3*, 130–153.

Morris, A. K., Hiebert, J., & Spitzer, S. M. (2009). Mathematical knowledge for teaching in planning and evaluating instruction: What can preservice teachers learn? *Journal for Research in Mathematics Education*, *40*(5), 491–529.

Murray, F. B. (1996). Beyond natural teaching: The case for professional education. In F. B. Murray (Ed.), *The teacher educator's handbook: Building a knowledge base for the preparation of teachers* (pp. 3–12). San Francisco, CA: Jossey-Bass.

National Council of Teachers of Mathematics. (1989). *Curriculum and evaluation standards for school mathematics*. Reston, VA: Author.

National Council of Teachers of Mathematics. (2000). *Principles and standards for school mathematics*. Reston, VA: Author.

National Governors Association Center for Best Practices & Council of Chief State School Officers. (2010). *Common core state standards for mathematics*. Washington, DC: Authors.

Improving Teachers' Assessment Literacy in Singapore Mathematics Classrooms: Authentic Assessment Task Design

Kim Koh and Olive Chapman

1 Introduction

As we are approaching the third decade of the twenty-first century, reforms-oriented professional development in classroom assessment for mathematics teachers has taken place in many education systems around the globe. This is especially so in high-performing education systems, which are characterized by their students' high-ranking performances in international assessments, such as the Program for International Student Assessment (PISA) and the Trends in International Mathematics and Science Study (TIMSS). One of the aims of PISA is to assess and compare the mathematical literacy of 15-year-old students across different education systems (Prenzel, 2013). Mathematical literacy denotes the ability to grasp the implications of many mathematical concepts, to reason and communicate mathematically, and to solve nonroutine, real-world problems effectively using a variety of mathematical methods (OECD, 2013). These competencies align well with the essential twenty-first century skills (e.g., critical thinking, creativity and innovation, problem-solving, communication, and collaboration) that are increasingly in demand in a complex, technologically connected, and fast-changing world (Koh, 2014).

With science, technology, engineering, and mathematics (STEM) fields of study and careers on the rise, mathematical literacy is deemed to prepare K-12 students for pursuing these interests. Although PISA claims to incorporate real-world tasks, students are only required to solve mathematical problems that are embedded within "camouflaged" open-ended tasks that require students to construct answers using a few words or short sentences. Many of the tasks students will face in their future workplaces and lives are much more complex than the PISA tasks. Therefore, there

K. Koh (✉) • O. Chapman
Werklund School of Education, University of Calgary, Calgary, AB, Canada
e-mail: khkoh@ucalgary.ca; chapman@ucalgary.ca

T. Leuders et al. (eds.), *Diagnostic Competence of Mathematics Teachers*, Mathematics Teacher Education 11, DOI 10.1007/978-3-319-66327-2_11

is a need for students in the day-to-day mathematics classrooms to be exposed to richer and authentic assessment tasks that enable them to demonstrate their mathematical understanding through complex problem solving in real-world contexts. Further, students should be provided with quality feedback in the process of learning mathematics. Taken together, these two reasons suggest that mathematics teachers need to be equipped with an adequate level of assessment literacy.

Teachers' assessment literacy is defined as teachers' sound knowledge and understanding of the principles and practices of assessment (Stiggins, 1991). Teachers' capacity to design, select, and use authentic assessment tasks to promote student learning is a key aspect of teachers' assessment literacy (Koh, Burke, Luke, Gong, & Tan, 2017). In this chapter, we report on a school-based, practice-oriented professional development approach to improving mathematics teachers' assessment literacy, specifically in designing and implementing authentic assessment tasks to promote students' mathematical literacy. We use the first author's work with Singaporean teachers as an example. We posit that teachers' active involvement in the design and implementation of mathematics authentic assessments has the potential to increase their diagnostic competence. Hoth et al. (2016) pointed out that "diagnostic competence of mathematics teachers is one special facet of teachers' professional competencies" (p. 43). Likewise, we deem that diagnostic competence is another important aspect of mathematics teachers' assessment literacy. Mathematics teachers' diagnostic competence refers to their capacity to identify students' preconceptions, misconceptions and errors, learning styles, language and cultural differences, interests, and motivation levels. Such a diagnostic competence enables teachers to provide quality feedback to students in the learning process, which in turn will lead to improvements in students' mathematics performance. Quality feedback is about the provision of timely and informative information to enable students to close the gap between the actual level and the reference level of their learning and achievement (Ramaprasad, 1983). Without proper feedback, the whole concept of authentic assessment contributing to students' learning is endangered. The chapter will begin with a contextual background of mathematics curriculum reforms in general and its impact on Singapore mathematics curriculum and teachers' assessment practices. We will then provide the rationale for using a school-based, practice-oriented professional development approach to improving the assessment literacy of elementary mathematics teachers. The contents of the professional development and the framework used to improve the teachers' competency in designing mathematics assessment tasks with high authentic intellectual quality will be included in the chapter. The effects of the professional development on the quality of teachers' mathematics assessment tasks and students' learning of mathematics will also be discussed. The chapter will end with some recommendations on the potential of using mathematics authentic assessment to increase mathematics teachers' diagnostic competence. Suggestions for future research are also included.

2 Mathematics Curriculum Reforms and Authentic Assessment

Curriculum reforms in mathematics are not new; however, they are increasingly important in a competitive global world that places greater emphasis on information technology as well as on STEM fields of study and careers. The literature on mathematics education has long advocated for a shift of focus from the drill and practice of basic mathematical concepts and procedural skills to students' active learning and understanding of complex mathematical concepts through nonroutine problem solving, mathematical thinking and reasoning, communication, and making connections to the real world (e.g., Chapman, 2013; Hiebert & Carpenter, 1992; Putnam, Lampert, & Peterson, 1990; Romberg, 2001; Schoenfeld, 1992). Learning mathematics with understanding or mathematical literacy has also been endorsed by the National Council of Teachers of Mathematics (NCTM) Standards in the US (NCTM, 1989) and the Cockcroft Report *Mathematics Counts* in the UK (Cockcroft, 1982). According to the NCTM standards (1989, 1991, 1995), the following five general goals are essential for students to become mathematically literate: (1) becoming a mathematical problem solver, (2) learning to reason mathematically, (3) learning to communicate mathematically, (4) learning to value mathematics, and (5) becoming confident of one's own ability. In addition, in the Principles and Standards for School Mathematics, the NCTM (2000) has called for teachers to adopt alternative forms of assessment that are aligned with these higher-order learning goals and that yield formative information about students' mathematical literacy. In essence, assessment tasks in mathematics classrooms need to be cognitively demanding and intellectually challenging to students so that they will engage in learning mathematics with understanding, which in turn leads to the development of their mathematical literacy. Further, assessment information should help teachers document and support student learning (i.e., formative assessment). This suggests that diagnostic competence is an important aspect of teachers' assessment literacy.

The Singapore Mathematics Curriculum Framework reflects the NCTM standards, which aim to develop students' mathematical literacy. It serves as a guidepost for the improvement of mathematics instruction and assessment in Singapore schools.

As can be seen in Fig. 1, mathematical problem solving is at the core of mathematics learning, and its development is dependent on five interwoven components: concepts, skills, processes, attitudes, and metacognition. This means to become mathematically literate, students do not only have to learn the concepts and procedural skills of mathematics, but also know how to use these skills to solve nonroutine problems through reasoning, communication, and making connections with the real world. In addition, students should be confident of their mathematical thinking and problem solving as well as be able to value mathematics and to self-regulate their own learning. This affective or noncognitive domain of learning outcomes is deemed to be increasingly important in today's educational contexts where teachers need to be cognizant of "the many ways in which student learning can unfold in the

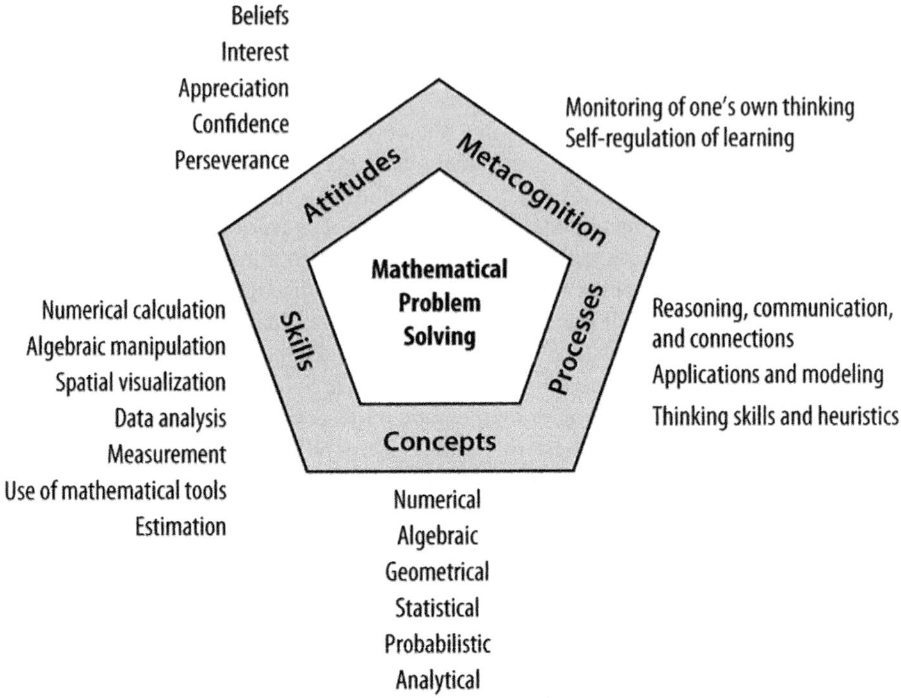

Fig. 1 Singapore Mathematics Curriculum Framework (MOE, 2006)

context of development, learning differences, language and cultural differences, and individual temperaments, interests, and approaches to learning" (Bransford, Darling-Hammond, & LePage, 2005, p. 1).

In line with the mathematics curriculum, Singaporean teachers have been urged to adopt alternative forms of assessment, which include authentic assessment and formative assessment. However, a lack of assessment literacy poses challenges to many mathematics teachers in the design, selection, and use of authentic assessment tasks. A common phenomenon is observed in other education systems. For example, research has shown that USA teachers' low level of assessment literacy impedes the implementation of high-quality performance assessments in mathematics (Borko, Mayfield, Marion, Flexer, & Cumbo, 1997; Silver, Mesa, Morris, Star, & Benken, 2009) despite a clarion call for assessment reforms. Fan (2002) and Koh and Luke (2009) found that most of the elementary mathematics teachers in Singapore schools were not familiar with performance-based tasks and that there was a misalignment between the desired mathematics curriculum and the enacted assessments in teachers' day-to-day classrooms. As such, they have called for the provision of systematic professional development in authentic assessment to help Singaporean teachers learn how to integrate this new form of assessment with the mathematics curriculum.

In short, teachers who implement mathematics curriculum reforms must change their assessment practices to include authentic assessments that enable richer demonstration and more holistic representation of what students know and can perform in mathematical and real-world contexts. Thus, there is a need for teacher professional learning or professional development approaches to focus on building mathematics teachers' capacity in designing and implementing classroom assessments that are well aligned with the objectives of reform-oriented mathematics curriculum.

3 Mathematics Teachers' Professional Learning in Assessment

The most common approach to help teachers has been to provide them with assessment resources to support classroom practices (Beesey, Clarke, Clarke, Stephens, & Sullivan, 1998). However, teachers are unlikely to learn how to use ready-made assessment tools and rubrics effectively on their own. As Webb (2009) noted, "Facilitating change in teachers' assessment practice is not so much a resource problem as it is a problem of … helping teachers develop a 'designers' eye' for selecting, adapting and designing tasks to assess student understanding" (p. 3). The works of others have also suggested the importance of educating teachers in the design and use of assessment tasks as a means of improving the quality of assessment tasks (Clarke, 1996; Koh, 2011a; Senk, Beckmann, & Thompson, 1997). According to Schoenfeld (2002), "standards-based reform appears to work when it is implemented as part of a coherent systemic effort in which curriculum, assessment, and professional development are aligned." (p. 17).

Using a collaborative professional development approach, the first author and her research team have worked with a group of Singaporean mathematics teachers who taught Grade 5 mathematics in authentic assessment task design. The 2-year intervention study contributes to our understanding of how a professional development approach focused on authentic assessment task design could impact the quality of teachers' assessment tasks and students' work in elementary school mathematics. Findings from the study also help us understand how teachers' active involvement in mathematics authentic assessment design might have potential to increase their diagnostic competence.

The elementary mathematics teachers were actively involved in a period of ongoing, sustained professional development to codesign authentic assessment tasks with their colleagues from the same school, subject, and grade level. Active learning and collective participation are two of the important features of effective professional development for teachers (Garet, Porter, Desimore, Birman, & Yoon, 2001). Research in mathematics education has pointed out that professional development which focuses on specific content and how students learn that content has greater positive effects on student learning than professional development which focuses on general pedagogy (Koh, 2014). As such, the foci of the professional development in this study were on teachers' mastery of both content knowledge (i.e., knowledge of the subject matter content in order to select tasks) and pedagogical content knowledge (i.e., knowledge of how students learn specific content in order to select tasks) so that changes in teachers' classroom practices would bring about positive student learning outcomes.

4 Professional Development in Authentic Assessment Task Design

The study was framed in a school-based professional learning community where teacher professional development activities were localized and contextualized in the teachers' mathematics classrooms. Participants consisted of 18 Grade 5 teachers from four schools selected at random from the same school system in Singapore. Ten teachers from two of the schools formed the intervention group and the other eight from the other two schools formed the comparison group. The intervention and comparison schools were matched based on two variables, namely, type of school and Ministry of Education's ranking. The schools were interested in the study so as to support their design and implementation of alternative forms of assessment based on the mathematics curriculum (i.e., the Pentagon Model in Fig. 1).

The participating teachers from the intervention group received ongoing, sustained professional development in mathematics authentic assessment task design and rubric development over the 2 years of the study while the teachers from the comparison group received only one workshop in authentic assessment at the end of each year. In authentic assessment initiatives, teachers' use of rubrics (i.e., a set of agreed-upon assessment criteria and standards) to enable them make defensible judgment of students' work is essential. Following are the key activities of professional development with the intervention group. First, the participating teachers engaged in identifying and stating learning goals for a unit of work. Second, they received both theoretical knowledge and practical training that helped them improve their understanding of the principles and features of authentic assessments and high-quality rubrics, as well as the Koh (2011b) criteria for authentic intellectual quality (AIQ) and their indicators in mathematics. The criteria and indicators aligned well with the desired learning outcomes in the new reform-oriented Singapore mathematics curriculum framework that they were required to implement. Third, the participating teachers were involved in codesigning mathematics authentic assessment tasks and associated rubrics with the help of the researcher and research assistants who included assessment specialists and mathematics content experts. The criteria of AIQ and their indicators (see Table 1) were used as guidelines for the teachers to codesign mathematics authentic assessment tasks, which were implemented in their teaching of mathematics.

Table 1 Criteria for judging the authentic intellectual quality of assessment tasks

Criteria	Examples of indicators
Depth of knowledge:	
Factual knowledge	Recognize mathematical terms; identify objects, patterns, or list properties; recall rules, formulae, algorithms, conventions of number, or symbolic representations
Procedural knowledge	Know how to carry out a set of steps; use a variety of computational procedures and tools
Advanced concepts	Make connections to other mathematical concepts and procedures; explain one or more mathematical relations
Knowledge criticism:	
Presentation of knowledge as a given	Accept or present ideas or information as a fixed body of facts; follow a set of preordained procedures
Comparing and contrasting information	Identify the similarities and differences in observations, data, or theorems; develop heuristics to identify, organize, classify, compare, and contrast data
Critiquing information	Comment on different mathematical solutions; make mathematical arguments
Knowledge manipulation:	
Reproduction	Recognize equivalents; perform a set of preordained algorithms; carry out computations
Organization, interpretation, analysis, evaluation, and synthesis of information	Interpret given mathematical equations, diagrams, tables, graphs, or charts; predict mathematical outcomes from the trends in the data
Application/problem solving	Apply mathematical concepts and procedures to solve nonroutine problems
Generation/construction of new knowledge	Come up with new proofs or solutions to a mathematical problem; generate mathematical procedures, strategies, or solutions to new problem situations
Extended communication	Elaborate on mathematical reasoning through arguments, prose, diagrams, sketches, drawings, or symbolic representations
Making connections to the real world beyond the classroom	Address a question, issue, concept, or problem that resembles one that they have encountered or are likely to encounter in daily life beyond the classroom

In addition to the practice-oriented professional development sessions, several monthly professional learning community meetings were held within each school where the participating teachers and the researcher addressed problems that have arisen from the implementation of the newly designed mathematics authentic assessments. The participating teachers' professional learning also included selecting and analyzing student work samples using the criteria for AIQ in moderation sessions. Such a highly contextualized learning activity supports the teachers in developing their ability to analyze students' mathematical thinking and to diagnose students' errors and misconceptions. This enables the teachers to develop their interpretive power, agency and autonomy in their professional judgment of students' work, and diagnostic competence (Koh, 2014; Nickerson & Masarik, 2010). Table 2 presents the criteria of AIQ and their indicators for teachers' judgment of students' work. As what you assess is what you get, the same AIQ criteria were used for both assessment tasks and students' work.

5 Impact on the Quality of Teachers' Assessment Tasks and Students' Work

The Singapore study examined the impact of the collaborative professional development approach to authentic assessment task design on the quality of mathematics teachers' assessment tasks and students' work through the use of a quasi-experimental, pre-post intervention design.

Data consisted of a random sample of teachers' assessment tasks (total 116 tasks) and related students' work (total of 712) collected at the beginning (Baseline), the end of year 1 (Phase I), and the end of year 2 (Phase II) of the professional development (intervention). Student work serves as a valid measure of changes in response to the intervention in this study because it has better instructional validity or sensitivity than test or exam scores (McClung, 1979; Popham, 2009). Having an immediate distance from the enactment of a particular lesson, student work is deemed to be the optimal indicator of student performance and learning (Ruiz-Primo, Shavelson, Hamilton, & Klein, 2002).

The teachers' assessment tasks and student work samples were blindly scored by the researcher, research assistants, and teachers using the criteria for AIQ. Each of the criteria employed a 4-point rating scale (ranging from 1 = no requirement/no demonstration to 4 = high requirement/high level). For example, factual knowledge was assessed using these scale points: 1 = no requirement in the assessment task, 2 = minimal requirement, 3 = moderate requirement, and 4 = high requirement. The interrater reliability for each of the criteria was determined using the percentage of exact agreement and kappa coefficient.

The results indicated that the intervention group of teachers increased their competence to design mathematics assessment tasks that were of high AIQ while the comparison group of teachers' tasks focused less on students' mathematical understanding, thinking, problem solving, and connections. The following are highlights of some of the key findings.

Table 2 Criteria for judging the authentic intellectual quality of students' work

Criteria	Examples of indicators
Depth of knowledge:	
Factual knowledge	Evidence of knowing mathematical terms, basic concepts, facts, properties, or principles; evidence of recitation or recognition of rules, formulae, algorithms, conventions of number, or symbolic representations
Procedural knowledge	Evidence of applying a set of steps; evidence of using a variety of computational procedures and tools
Advanced concepts	Evidence of making connections to other mathematical concepts and procedures; evidence of an explanation of one or more mathematical relations
Knowledge criticism:	
Presentation of knowledge as a given	No evidence of critiquing or questioning of ideas or information; evidence of performing a set of preordained procedures
Comparing and contrasting information	Evidence of an identification of similarities and differences in observations, data, or theorems; evidence of performing heuristics to identify, organize, classify, compare, and contrast data
Critiquing information	Evidence of critiquing different mathematical solutions; evidence of making mathematical arguments
Knowledge manipulation:	
Reproduction	Evidence of recognizing equivalents; evidence of performing a set of preordained algorithms; evidence of carrying out computations
Organization, interpretation, analysis, evaluation, and synthesis of information	Evidence of interpreting given mathematical equations, diagrams, tables, graphs, or charts; evidence of predicting mathematical outcomes from the trends in the data
Application/problem solving	Evidence of applying mathematical concepts and procedures to solve nonroutine problems
Generation/construction of new knowledge	Evidence of creating new proofs or solutions to a mathematical problem; evidence of generalizing mathematical procedures, strategies, or solutions to new problem situations
Extended communication	Ability to elaborate on mathematical reasoning through arguments, prose, diagrams, sketches, drawings, or symbolic representations
Making connections to the real world beyond the classroom	Making connections between responses to task questions and the world beyond the classroom; relating mathematical knowledge and skills to real world problems or personal experiences

5.1 Quality of Mathematics Assessment Tasks

Teachers from the intervention schools focused less on assessing students' *factual and procedural knowledge*, *presentation of knowledge as a given*, and *knowledge reproduction*. This means they placed greater emphasis on the following criteria in their mathematics authentic assessment task design: *understanding of advanced concepts*, *comparing and contrasting knowledge*, *critique of knowledge*, *organization, interpretation, analysis, synthesis, and evaluation*, *problem solving*, *generation of new knowledge*, *extended communication*, and *making connections to the real world*. These criteria reflect mathematical literacy and the essential twenty-first century competencies as discussed in the preceding sections of this chapter. After their active participation in the professional development sessions and the professional learning communities, the teachers became more assessment literate. Based on anecdotal evidence, they started to aim for mathematics assessment tasks that were of high AIQ, which enabled the assessment of students' mathematical literacy and twenty-first century competencies. Their professional conversations during the collaborative design of mathematics authentic assessment tasks indicate that they devoted considerable attention to the potential of using authentic tasks and associated rubrics to identify and understand students' preconceptions, learning styles, language and cultural differences, interests, and motivation levels.

5.2 Quality of Student Work in Mathematics

What you assess is what you get! Similar to the assessment tasks, student work samples collected from both the intervention and comparison schools were analyzed using the criteria of AIQ. Students' work in intervention schools demonstrated less *presentation of knowledge as a given* and *knowledge reproduction*. There was a significant increase on the following: *comparing and contrasting knowledge*, *organization, interpretation, analysis, synthesis and evaluation*, and *problem solving*. In contrast, there was a significant decrease on *problem solving* and *extended communication* in students' work collected from the comparison schools.

Through reform-oriented professional development activities, the teachers in the two intervention schools increased their competence to design mathematics assessment tasks that were of high AIQ. As a result, their students were able to demonstrate better performances on the tasks assigned by them. Before intervention, most of the assessment tasks given to students contained routine problem-solving questions, which only required students to show the procedures of arriving at the correct answers. After intervention, there was a significant improvement of the quality of the mathematics assessment tasks. Most of the tasks focused on students' nonroutine problem solving in real-world contexts.

Appendix A presents an example of a Grade 5 mathematics authentic assessment task codesigned by the teachers in one of the intervention schools. The task included a real-world scenario in which students were asked to compare and contrast prices of different models of camera from an advertisement and show the differences of prices paid by two different methods, that is, cash versus installments. In order to be successful on completing the authentic task, students were required to demonstrate their mathematical thinking and reasoning (i.e., making their thinking visible), non-routine problem solving, communication (i.e., argumentation) and making connections to real world (i.e., decision making on the purchase of a camera after comparing and contrasting the different types of camera and their prices). In addition, students should be confident of their mathematical thinking and problem solving as well as be able to value the learning of the concept of percentage and to self-regulate their own learning (i.e., using the criteria and standards on a rubric given by the teacher to self-assess and monitor learning progression). Such an authentic task provides ample opportunity for teachers to diagnose not only students' cognitive processes but also students' affect-motivational skills (Hoth et al., 2016).

6 Discussion and Recommendations

The findings from the Singapore study indicate that mathematics teachers' assessment literacy, specifically their competence in the design and implementation of authentic assessment tasks can be significantly increased when they take part in school-based, practice-oriented professional development activities that are localized and contextualized in their classrooms. Instead of attending traditional assessment workshops and adopting prescribed, ready-made assessment tools, the teachers in the intervention schools codesigned authentic assessment tasks that were aligned with the local mathematics curriculum and that were relevant to the local school context. This is similar to the findings on the use of highly adaptive videos in teacher professional learning. Highly adaptive video clips were selected from the participating teachers' classrooms and hence the teachers could see the relevance between the contents of professional development and the needs in their own classrooms (Koh, 2015).

Teachers' active and collective participation in mathematics authentic assessment task design enabled them to develop the "designers' eyes" so that they were able to design and use authentic tasks that assess and promote students' mathematical literacy. As a result, it improved the quality of students' work in mathematics learning. The participating teachers' active involvement in codesigning mathematics authentic assessment tasks as well as in the analysis and moderation of students' work using the criteria of AIQ also increased their understanding of students' mathematical thinking and reasoning, preconceptions, errors, and misconceptions. This provided them with the opportunity to reflect on the mathematics curriculum and to use the assessment information to adjust their own lesson plans and instructional practice as well as to provide students with quality feedback. Such a formative assessment practice is key to improving the quality of student learning in the daily mathematics classroom. It is also well aligned with the Principles and Standards for

School Mathematics (NCTM, 2000), which has called for mathematics teachers to adopt formative assessment practice.

In view of the importance of improving mathematics teachers' diagnostic competence, we have made three recommendations as follows. First, mathematics authentic assessment tasks and associated rubrics should include clearly defined learning progressions. The learning progressions map out a specific sequence of knowledge, skills, and dispositions that students are expected to learn and master as they progress through the continuum. Clearly defined learning progressions will enhance teachers' diagnostic competence, which results in quality feedback that helps students close the gap of their learning and achievement. Second, mathematics teachers' professional judgment of students' work in social moderation sessions can serve as a catalyst for improving their diagnostic competence. The professional conversations that take place during moderation sessions will help improve teachers' capacity to interpret and understand students' preconceptions, errors and misconceptions, learning styles, language and cultural differences, interests, and motivational levels. Third, diagnostic competence is a key facet of teachers' assessment literacy and hence professional development in mathematics authentic assessment should include learning opportunities for increasing teachers' diagnostic competence.

Our work with elementary mathematics teachers supports Webb's (2009) argument for the importance of developing teachers' expertise in classroom assessment so that teachers have better conceptions of and confidence in using alternative forms of assessment to facilitate the implementation of new mathematics curricula. It is important for future research to examine the change in elementary mathematics teachers' conceptions of and confidence in using authentic assessment tasks to support student learning of mathematics with understanding (i.e., mathematical literacy) after their participation in professional development with a focus on mathematics authentic assessment task design. The findings of our study also indicate significant potential for using the criteria of AIQ as the guideposts for improving the quality of teachers' assessment tasks in K-12 mathematics teaching and learning. The criteria and indicators of AIQ are generic and hence they can be applied to different content areas, assessment tasks, and grade levels. For example, the first author of this chapter has used the same criteria to work with a group of Canadian elementary mathematics teachers in designing authentic assessment tasks to assess Grade 6 students' understanding of geometry and measurement.

One of the limitations of the Singapore study was a lack of direct measurement of teachers' assessment literacy. Instead, the quality of teachers' assessment tasks was used as a proxy measurement of teachers' assessment literacy. However, while the ultimate measure of change in teacher quality through professional development is improved student learning and mathematical literacy, we deemed that students' work serves as an effective measure of change in teacher quality, that is, teacher assessment literacy in the context of this study. Finally, due to time constraints of the participating teachers and a need to minimize instructional disruptions, we did not have any opportunities to conduct classroom observations of the implementation of the authentic assessment tasks in their classrooms. In future research, this important component will need to be negotiated with teacher participants. This will enable the researchers to make meaningful inferences about the enactment of authentic assessment practices in actual mathematics instruction and to understand the impact of mathematics authentic assessments on teachers' instructional practices.

Appendix A

S13/P5/MA/-/B/E5/T7/14/3

Primary School
P5 Mathematics – Percentage

P : 2/3
C : 3/3
A : 1/3
(6)/9

Name: ____________________ (   )

Class: 5_ ____________

Date: 23rd August 2007

Study the advertisement given. The advertisement given to you shows 6 different models of digital camera and their prices.

References

Beesey, C., Clarke, B. A., Clarke, D. M., Stephens, M., & Sullivan, P. (1998). *Effective assessment for mathematics*. Carlton, VIC: Board of Studies/Addison Wesley Longman.

Borko, H., Mayfield, V., Marion, S., Flexer, R., & Cumbo, K. (1997). Teachers' developing ideas and practices about mathematics performance assessment: Success, stumbling blocks, and implications for professional development. *Teaching and Teacher Education, 13*(3), 259–278.

Bransford, J., Darling-Hammond, L., & LePage, P. (2005). Introduction. In L. Darling-Hammond & J. Bransford (Eds.), *Preparing teachers for a changing world: What teachers should learn and be able to do* (pp. 1–39). San Francisco, CA: Jossey-Bass.

Chapman, O. (2013). Engaging children in learner-focused mathematical tasks. *Plenary presentation at the 13th Biannual International Symposium for Elementary Mathematics Teaching* [SEMT-13], Charles University, Prague.

Clarke, D. (1996). Assessment. In A. J. Bishop, K. Clements, C. Keitel, J. Kilpatrick, & C. Laborde (Eds.), *International handbook of mathematics education* (Vol. 1, pp. 327–370). Dordrecht, The Netherlands: Kluwer Academic.

Cockcroft, W. H. (1982). *Mathematics counts. The Cockcroft report*. London: Her Majesty's Stationery Office.

Fan, L. (2002). In-service training in alternative assessment with Singapore mathematics teachers. *The Mathematics Educator, 6*(2), 77–94.

Garet, M. S., Porter, A. C., Desimore, L., Birman, B. F., & Yoon, K. S. (2001). What makes professional development effective? Results from a national sample of teachers. *American Educational Research Journal, 38*(4), 915–945.

Hiebert, J., & Carpenter, T. P. (1992). Learning and teaching with understanding. In D. A. Grouws (Ed.), *Handbook of research on mathematics teaching and learning* (pp. 65–97). New York, NY: Macmillan.

Hoth, J., Döhrmann, M., Kaider, G., Busse, A., König, J., & Blömeke, S. (2016). Diagnostic competence of primary school mathematics teachers' during classroom situations. *ZDM Mathematics Education, 48*, 41–53.

Koh, K. (2011a). Improving teachers' assessment literacy through professional development. *Teaching Education, 22*(3), 255–276.

Koh, K. (2011b). *Improving teachers' assessment literacy*. Singapore: Prentice Hall.

Koh, K. (2014). *Authentic assessment, teacher judgment and moderation in a context of high accountability*. In C. Wyatt-Smith, V. Klenowski, & P. Colbert (Eds.), *Designing assessment for quality learning* (Vol. 1, pp. 249–264). Dordrecht, Netherlands: Springer.

Koh, K. (2015). The use of video technology in pre-service teacher education and inservice teacher professional development. In S. F. Ng (Ed.), *Cases of mathematics professional development in east asian countries* (pp. 229–247). Singapore: Springer.

Koh, K., & Luke, A. (2009). Authentic and conventional assessment in Singapore schools: An empirical study of teacher assignments and student work. *Assessment in Education: Principles, Policy & Practice, 16*(3), 291–318.

Koh, K., Burke, L. E. C., Luke, A., Gong, W. G., & Tan, C. (2017). Developing the assessment literacy of teachers in Chinese language classrooms: A focus on assessment task design. *Language Teaching Research., doi: 10.1177/1362168816684366*

McClung, M. S. (1979). Competency testing programs: Legal and educational issues. *Fordham Law Review, 47*, 652–711.

Ministry of Education. (2006). *Mathematics syllabus primary*. Singapore: Curriculum Planning and Development Division, Ministry of Education.

National Council of Teachers of Mathematics. (1989). *Curriculum and evaluation standards for school mathematics*. Reston, VA: Author.

National Council of Teachers of Mathematics. (2000). *Principles and standards for school mathematics*. Reston, VA: National Council of Teachers of Mathematics.

National Council of Teachers of Mathematics. (1991). *Professional standards for teaching mathematics*. Reston, VA: Author.

National Council of Teachers of Mathematics. (1995). *Assessment standards for school mathematics*. Reston, VA: Author.

Nickerson, S. D., & Masarik, D. K. (2010). Assessing teachers' developing interpretive power: Analysing student thinking. *Mathematics Teacher Education and Development, 12*(1), 19–29.

OECD. (2013). *PISA 2012 Assessment and analytic framework: Mathematics, reading, science, problem-solving, and financial literacy*. Paris: PISA, OECD Publishing.

Popham, W. J. (2009). Assessment literacy for teachers: Faddish or fundamental*? Theory Into Practice*, *48*, 4–11.

Prenzel, M. (2013). Introduction: Research on PISA, with PISA, and for PISA. In M. Prenzel, M. Kobarg, K. Schöps, & S. Rönnebeck (Eds.), *Research on PISA*. Dordrecht, Netherlands: Springer.

Putnam, R. T., Lampert, M., & Peterson, P. L. (1990). Alternative perspectives on knowing mathematics in elementary schools. In C. Cazden (Ed.), *Review of research in education* (Vol. 16, pp. 57–152). Washington, DC: American Educational Research Association.

Ramaprasad, A. (1983). On the definition of feedback. *Behavioural Science*, *28*, 4–13.

Romberg, T. A. (2001). Understanding the standards-based reform movement in school mathematics. *The Mathematics Educator*, *6*(1), 17–21.

Ruiz-Primo, M. A., Shavelson, R. J., Hamilton, L., & Klein, S. (2002). On the evaluation of systemic science reform: Searching for instructional sensitivity. *Journal of Research in Science Teaching*, *39*(5), 369–393.

Schoenfeld, A. H. (2002). Making mathematics work for all children: Issues of standards, testing, and equity. *Educational Researcher*, *31*(1), 13–25.

Schoenfeld, A. H. (1992). Learning to think mathematically: Problem solving, metacognition, and sense-making in mathematics. In D. Grouws (Ed.), *Handbook for Research on Mathematics Teaching and Learning (pp. 334–370). New York: MacMillan.*

Senk, S. L., Beckmann, C. B., & Thompson, D. R. (1997). Assessment and grading in high school mathematics classrooms. *Journal of Research in Mathematics Education*, *28*, 187–215.

Silver, E. A., Mesa, V. M., Morris, K. A., Star, J. R., & Benken, B. M. (2009). Teaching mathematics for understanding: An analysis of lessons submitted by teachers seeking NBPTS certification. *American Educational Research Journal*, *46*(2), 501–531.

Stiggins, R. J. (1991). Assessment literacy. *Phi Delta Kappan*, *72*(7), 534–539.

Webb, D. C. (2009). Designing professional development for assessment. *Educational Designer, 1*(2), 1–26. Retrieved from http://www.educationaldesigner.org/ed/volume1/issue2/article6

Developing Prospective Teachers' Ability to Diagnose Evidence of Student Thinking: Replicating a Classroom Intervention

Christine M. Phelps-Gregory and Sandy M. Spitzer

1 Introduction

In order for teachers to improve over time, they must be proficient at collecting and analyzing evidence of student thinking and learning (Hiebert, Morris, Berk, & Jansen, 2007). This specific type of diagnostic competence, which focuses on diagnosing student learning with the specific goal of studying and improving teaching, can be improved through interventions in teacher education (see, e.g., Spitzer, Phelps, Beyers, Johnson, & Sieminski, 2011). In this chapter, we discuss the findings of previous interventions aimed at helping prospective teachers (PTs) learn to analyze student thinking. Then, we present a replication study using a classroom intervention to teach prospective elementary teachers ($N = 23$) to identify and evaluate evidence of student understanding. Results of this study and previous work show that diagnostic competence is a skill that is teachable through interventions. After the intervention described in this chapter, participants performed better on a measure of diagnostic competence. In particular, they improved their ability to distinguish evidence of student thinking from nonevidence, such as a teacher's lecture. They were also more likely to recognize that students' procedural work cannot be used to diagnose conceptual understanding. Results will be used to suggest key features of interventions to improve diagnostic competence.

Research suggests that in order for teachers to become expert practitioners, they must engage in skillful, systematic reflection which targets student learning (Feiman-Nemser, 2001; Hammerness, Darling-Hammond, & Bransford, 2005).

C.M. Phelps-Gregory (✉)
Central Michigan University, Mount Pleasant, MI, USA
e-mail: phelp1cm@cmich.edu

S.M. Spitzer
Towson University, Towson, MD, USA

T. Leuders et al. (eds.), *Diagnostic Competence of Mathematics Teachers*, Mathematics Teacher Education 11, DOI 10.1007/978-3-319-66327-2_12

When teachers systematically study the effects of their instruction on students' achievement of mathematical learning goals, they have the potential to improve their teaching effectiveness over time (Hiebert et al. 2007). In particular, diagnosing student thinking allows teachers to make changes to their instruction which directly target student learning (Berk & Hiebert, 2009; Hiebert et al., 2007; Santagata & Yeh, 2014). To do this, teachers must attend to the sometimes small but mathematically important details in student thinking which reveal those students' achievement of mathematical learning goals (Morris, Hiebert, & Spitzer, 2009). Thus, teachers' ability to make correct and appropriate judgments about student understanding, sometimes termed diagnostic competence (Leuders, Dörfler, Leuders, & Philipp, 2018 chapter "Diagnostic Competence of Mathematics Teachers: Unpacking a Complex Construct" in this book), is a necessary skill for the lifelong learning envisioned by education reformers (e.g., Feiman-Nemser, 2001).

Furthermore, emerging evidence exists that teachers who are explicitly taught how to elicit and diagnose student thinking for the purposes of improving their own instruction may also perform better on other tasks of teaching. For example, Santagata and Yeh (2014) found that when prospective teachers (PTs) were explicitly trained in diagnostic competence, their use of rich questioning to elicit and incorporate student thinking during classroom lessons improved (compared to PTs who had not been educated in diagnostic competence). Diagnostic competence is also closely related to the construct of teacher noticing (see, e.g., Jacobs, Lamb, & Philipp, 2010; Sherin, Jacobs, & Philipp, 2011). Researchers studying teacher noticing have found that when teachers can more skillfully attend to and interpret student thinking, they are also better able to respond to that thinking in the moment (Jacobs et al., 2010; Schoenfeld, 2011).

Teachers might have many differing reasons to judge students' mathematical understanding (Philipp & Leuders, 2014). For example, teachers must appropriately diagnose student thinking in order to select and sequence student solutions for a whole-class discussion (see Meikle, 2014). There is also substantial overlap between diagnostic competence and formative assessment, where student thinking is diagnosed for a specific purpose.

In this chapter, we are specifically interested in teachers' ability to diagnose their students' understanding of mathematics for the specific goal of studying and improving teaching. Here, this aspect of diagnostic competence is conceptualized as PTs' ability to determine the extent to which student responses (e.g., student talk, written work, and non-verbal actions in a whole-class setting) provide evidence that the student has achieved (or not achieved) a specified mathematical learning goal. For the remainder of this chapter, we use such a conceptualization of this aspect of diagnostic competence, which we term "diagnosing evidence of student thinking."

2 Using Interventions to Improve the Prospective Teachers' Ability to Diagnose Evidence of Student Thinking

An emerging set of research has investigated classroom interventions for improving PTs' ability to diagnose evidence of student thinking. In particular, these studies have tried to improve the correctness of teachers' diagnoses, that is, the "correspondence between teachers' diagnostic judgments and students' actual achievement" (Leuders et al., 2018 chapter "Diagnostic Competence of Mathematics Teachers: Unpacking a Complex Construct" in this book). The emerging consensus from such research is that diagnostic competence is a skill that can be taught to PTs (e.g., Santagata, Zannoni, & Stigler, 2007; Spitzer et al., 2011). This is promising given that, unlike practicing teachers, PTs have limited access to authentic classroom situations involving student thinking; the context of teacher education is necessarily somewhat removed from the classroom and such situations must appear through simulations or representations. Looking across multiple studies, we can generalize about the kinds of skills that an intervention can teach as well as the key features of successful interventions.

2.1 What Skills Can Be Learned by Prospective Teachers?

Multiple authors have designed and tested interventions to improve PTs' ability to diagnose evidence of student thinking, and they have operationalized this construct differently and focused on different skills. However, some commonalities can be found. One common finding among many studies is that PTs can learn to analyze lessons at a deeper or more meaningful level after an intervention (Alsawaie & Alghazo, 2010; Santagata & Angelici, 2010; Santagata & Guarino, 2011; Santagata et al., 2007; Spitzer & Phelps, 2011; Spitzer et al., 2011; Stockero, 2008). This depth of analysis has been measured in different ways, but always includes a focus on the mathematically important details of student responses and their implications. Across studies, it was found that after interventions, PTs perform better at choosing mathematically significant events in lessons, making more detailed and specific claims, and linking those claims to evidence. In particular, PTs' ability to interpret student responses has improved in several interventions. For example, Alsawaie and Alghazo (2010) and Stockero (2008) both found that after an intervention, PTs were better able to link classroom events with pedagogical principles and made more inferences about what students might understand (rather than making primarily snap judgments and descriptive comments).

Another way in which diagnostic competence has been measured, and in which PTs' skills have improved after an intervention, is the extent to which PTs take a "tentative" or "critical" stance in their diagnoses of student thinking (Bartell, Webel, Bowen, & Dyson, 2013; Santagata et al., 2007; Spitzer et al., 2011; Stockero, 2008). This indicates that after interventions, PTs shift from automatically assuming that students learn whatever teachers say to a disposition of seeking evidence of student thinking. For example, Stockero (2008) found that PTs became less certain, and more nuanced, in their judgments about the success of lessons after an intervention. On the other hand, it has also been found that PTs might become overly critical of evidence after an intervention. For example, both Spitzer et al. (2011) and Bartell et al. (2013) found that on a posttest, PTs were more likely to claim that a student had no understanding of a particular math content (rather than realizing that the evidence was insufficient for making such a claim).

Finally, interventions have been able to improve PTs' ability to suggest alternatives or revisions to a lesson that would support student learning and that are based on diagnosed student thinking (Santagata & Angelici, 2010; Santagata & Guarino, 2011; Santagata et al., 2007; Spitzer & Phelps, 2011). For example, Santagata and Angelici (2010), who asked PTs to revise a lesson to better support student learning, found that PTs who had experienced an intervention were more likely than a control group to link their revisions to their findings about student thinking. Similarly, Spitzer and Phelps (2011) found that after PTs participated in an online discussion board intended to improve their diagnostic competence, their lesson revisions were more related to student thinking about the learning goal of the lesson (compared with their initial revisions, which primarily targeted surface features or behavior issues). These results indicate that PTs can learn to use their diagnostic competence to improve their teaching over time. However, the conditions that promote such a use of diagnostic competence are not sufficiently well known.

2.2 *What Key Features Have Been Included in Successful Interventions?*

Again, because many different authors have studied diagnostic competence, the features of their interventions have been varied. One feature of interventions found in the literature was the use of representations of practice for PTs to analyze. These have taken the form of videos (including both edited lesson samples and full class sessions) and written lesson transcripts. Both of these representations of practice have been used successfully. Although it is difficult to compare results across studies using different measures and definitions of diagnostic competence, the evidence does not currently suggest that one form or the other is more effective. Several authors have developed specific criteria for deciding which kinds of lessons will be most useful for PTs. For example, most interventions (e.g., Alsawaie & Alghazo, 2010; Bartell et al., 2013; Spitzer & Phelps, 2011) have included videos or transcripts which include both evidence of student thinking (either clear misconceptions or conceptual explanations) and segments in which student thinking is unclear or non-revealing. This aligns with the recommendations of Sherin, Linsenmeier, and van Es (2009) who note that videos including "windows into student thinking" (p. 215) have been most effective in their work with practicing teachers.

Another key feature which appears across multiple interventions is the use of explicit frameworks for analysis (e.g., Alsawaie & Alghazo, 2010; Bartell et al., 2013; Santagata & Yeh, 2014; Santagata et al., 2007; Spitzer et al., 2011; Star & Strickland, 2008). Nearly all interventions have included at least some explicit instructor work to direct PTs' attention to specific features of student thinking. Star and Strickland (2008) argue for the importance of such frameworks, noting that "the framework appeared to be instrumental in beginning to direct preservice teachers' attention away from more superficial features of classrooms and toward aspects that are likely more critical in terms of mathematics teaching and learning" (p. 124).

2.3 *What Further Research Is Needed?*

While results of initial studies are promising, the field has further to go to understand the impact of interventions on PTs' diagnostic competence. For example, one feature of interventions which has been highly variable in the literature is their length. Studies range from the discussion board intervention used by Spitzer and Phelps (2011), which included only outside-of-class activities occurring over about a 2 week period, to the Learning to Learn From Teaching course described by Santagata & Yeh (2014), in which the entire semester counts as the intervention. Similarly, Star and Strickland (2008) considered an entire methods course as their intervention. Promising results seem to have come from both shorter and longer interventions, suggesting further research is needed to investigate the value of increased time on task for learning diagnostic competence. In addition, no work has looked at the long-term effects, particularly the effects on PTs' future teaching, of such interventions.

Finally, mathematics education researchers have argued for the importance of replication in education research (Adler, Ball, Krainer, Lin, & Novotna, 2005). No such replication work has yet been conducted on the effects of interventions to improve PTs' diagnostic competence abilities. The need for replication work is particularly important because previous work in this area has been primarily quasi-experimental, since assigning PTs randomly to interventions is difficult and ethically problematic. While quasi-experimental work can provide valuable insights, it is also important for researchers to consider threats to internal validity. Replication work can help us better understand the results of quasi-experimental work.

2.4 *Conceptualizing Diagnostic Competence*

In this chapter, we report on a replication study of Spitzer et al. (2011). Spitzer et al. (2011) used a quasi-experimental design, with a pre- and posttest but no control group, to examine the effects of an intervention on specific aspects of PTs' diagnostic competence. More details on their methodology are provided below. In this study, the authors investigated whether the intervention could help PTs take a critical stance as to whether potentially diagnostic information truly reveals students' understanding of mathematical ideas (that is, their ability to diagnose evidence of student thinking).

Primary findings of Spitzer et al. (2011) indicate that, after the intervention, PTs demonstrated stronger abilities to diagnose evidence of student thinking in some contexts: For example, they were more likely to recognize that only student work relevant to the learning goal could be used to diagnose student understanding. However, PTs did not improve in their ability to diagnose evidence of students' conceptual understanding and distinguish conceptual understanding from procedural fluency.

3 Previous Findings from Spitzer et al. (2011)

The Spitzer et al. (2011) study used a quasi-experimental pretest/posttest design to investigate an intervention aimed at improving PTs' ability to diagnose and evaluate evidence of student achievement of specified mathematical learning goals. Participants included prospective elementary teachers ($N = 160$) enrolled in a mathematics content course at a large, public university in the Mid-Atlantic region of the United States. The first in a sequence of three such required courses, this semester-long course focused on PTs' conceptual understanding of the base-10 number system and the meaning underlying common computational algorithms.

PTs completed a pre- and posttest and participated in an in-class intervention. The pretest was administered to measure participants' entering ability to diagnose evidence of student thinking. The pretest asked PTs to read and respond to a transcript of a classroom lesson. The learning goal of the lesson, clearly stated for PTs, was "Students will understand the key concepts involved in the common denominator strategy for comparing fractions." The transcript had six sections (including a teacher explanation with students nodding as well as student responses which were either irrelevant to the learning goal or contained strictly procedural mathematical work). For each section, PTs were asked to diagnose what they could tell about students' understanding of the learning goal in each section. It is important to note that the transcript was designed to provide no evidence of student understanding; therefore the best PT response would note that there was not enough evidence to diagnose student understanding. The posttest (completed approximately 3 weeks later) differed in superficial ways from the pretest (e.g., different student names and different order of activities, but the same kinds of evidence and mathematical content).

Between the pre- and posttests, PTs participated in an intervention composed of two 75 min lessons and associated homework assignments. First, PTs read a transcript and responded individually in writing to evaluate student conceptual understanding of a stated learning goal. This task was similar in structure to the pre- and posttest but included different mathematics content (focused on base-10 numbers).

Although the transcript did not contain any significant evidence of student achievement of the goal, most PTs responded with claims that students had learned this goal. During class time, instructors told PTs that the students in the lesson sample had performed poorly on a quiz the next day, and instructed PTs to return to the transcript to reconsider and revise their responses. PTs then discussed the transcript in small groups and participated in a large-group discussion in which some principles for analysis were discussed explicitly by the instructor. In the second lesson of the intervention, PTs completed a card-sort task, again focused around base-10 numbers, in which they matched learning goals to student responses showing evidence of achieving those goals. (This task was quite different in structure than the pre- or posttest.) Another large-group discussion highlighted the importance of looking for student responses which are both relevant to the learning goal and revealing of student thinking around that goal.

In order to allow for comparison, Spitzer et al. (2011) coded PTs' responses on each section of the pre- and posttest as a diagnosis of no evidence of student achievement of the learning goal (coded as 0), a diagnosis of some evidence (coded as 1), or a diagnosis of a lot of evidence (coded as 2). Since by design the transcripts contained no evidence of achievement of the learning goal, the best response for every section was a 0 and the lower mean responses indicated a better analysis.

Results from the posttest indicate that PTs improved their evaluations of evidence consisting primarily of teacher talk, with the mean score decreasing from 1.34 on the pretest to 0.99 on the posttest ($p < 0.001$). PTs also improved in their ability to analyze student responses which were irrelevant to the learning goal, with the mean responses on all three of these sections significantly improving (see Spitzer et al., 2011 for more details). However, PTs continued to take students' demonstrated procedural fluency with an algorithm as evidence of conceptual understanding of the meaning underlying that algorithm, with no significant changes in their mean ratings of the section which provided detailed evidence of procedure. In addition, some PTs also demonstrated an overly critical stance toward evidence on the posttest. Instead of realizing that additional evidence was needed to judge student thinking, these PTs responded that the evidence suggested that students did not understand the ideas of the lesson.

In the study described above, Spitzer et al. (2011) used a quasi-experimental pre- and posttest intervention design. In their limitations section, the authors address threats to internal validity stemming from the lack of a control group. However, one additional method to verify and strengthen results is replication. Thus, the primary goal in the study reported here was to replicate the methodology of the Spitzer et al. (2011) study in order to confirm and extend its findings. Replication is a necessary but under-utilized tool in the search for generalizable knowledge about the processes of education (Makel & Plucker, 2014). We enacted a conceptual replication (Schmidt, 2009), in which we duplicated the methods used in nearly every way, with adjustments (described below) to clarify procedure and to improve the intervention. The goal of a conceptual replication is to "systematically chang[e] individual facets of the original study to better understand its nature" (Makel & Plucker, 2014, p. 2).

4 Methods

The study described here included three phases: a pretest, an in-class intervention, and a posttest after the intervention. Both the pre- and posttests involved asking participants to read a written transcript and analyze it for evidence that could help them diagnose students' mathematical understanding of the learning goal. The intervention focused specifically on improving PTs' abilities to diagnose such evidence and was designed based on findings from previous research.

4.1 Participants

For this replication study, participants ($N = 23$) were enrolled in a course for prospective elementary and middle school (grades K-7) teachers at a large mid-western university in the United States. The course was a mixed content and methods course, focusing both on mathematics and on how to teach mathematics (pedagogy). In particular, the course focused on studying student thinking in mathematics and was intended for math majors and minors who might go on to teach middle school or become math specialists. PTs were taught to analyze cognitive demand, write learning goals, analyze evidence to diagnose learning, hypothesize about the links between teaching and learning, and revise lessons to improve student learning. The intervention presented here was conducted in weeks 3 and 4 of a 16 week semester. This was after PTs had learned about cognitive demand and learning goals but before they had learned any type of diagnostic skills. It is possible PTs had some prior diagnostic experience from classes such as their methods course for teaching reading. However, the mathematics methods course at this university focuses very little attention on diagnosis and the PTs had no other formal experience with this skill. This aligns with the methodology of Spitzer et al. (2011), whose participants had similarly received no explicit training on diagnosing evidence of student thinking before the study.

It is important to note here that the PTs in this study were mathematics majors and minors, in contrast with the participants in Spitzer et al.'s (2011) study. Those participants were general elementary education majors with potentially weaker mathematics backgrounds. Although they potentially knew more mathematics, the PTs in this study did not have more training in diagnosing evidence of student thinking in mathematics. Implications of this difference will be discussed below.

4.2 Measures and Analysis

PTs completed a transcript analysis task for both the pre- and posttest. The pre- and posttests were identical to each other (to aid in comparison). PTs were asked to read a written transcript of a lesson on ordering fractions. The learning goal for the lesson was displayed prominently at the top of both the transcript and the analysis task and stated, "The goal for this lesson was for students to understand the key concepts involved in the common denominator strategy for comparing fractions." Notably, this is a conceptual learning goal which specifies that students should understand a strategy, rather than just be able to carry it out. The transcript had six sections in total. The six sections of the transcript included two sections focused on teacher explanation with non-revealing responses (e.g., "Several students say 'Oh, I see'"), two sections where students gave procedural responses (even though the learning goal was conceptual), and two sections where students gave responses that were irrelevant to the learning goal (for example, drawing a picture to compare fractions, even though the picture did not include any attention to the ideas of the common denominator strategy). An excerpt from the transcript, used for both the pre- and posttests, is given below:

Ms. Green [the teacher]:	"Let's just look at 1/4 and 3/4. Which fraction do you think is larger?"
Jose:	"3/4"
Ms. Green:	"Jose, how do you know 3/4 is bigger than 1/4?"
Jose:	"Because 3 is a bigger number than 1."

Note that even though Jose gives the correct answer, his response does not reveal the achievement of the learning goal, since it does not include any attention to the ideas of the common denominator strategy. The transcript task used for the pre- and posttest came from Spitzer et al. (2011) but was modified so there were two of each type of section (to better allow for comparisons between PTs' ability on each type of evidence).

To align with Spitzer et al.'s (2011) coding scheme, on each section of the pre- and posttests, PTs were asked to evaluate if they could tell anything based on student understanding. That is, they were asked to circle that they "Can't tell" (a diagnosis of no evidence of student achievement, coded as 0), "Can tell a little" (a diagnosis of some evidence, coded as (1), or "Can tell a lot" (a diagnosis of a lot of evidence, coded as (2). None of the sections of the transcript revealed evidence of student understanding. Thus, the highest quality response to this transcript would be to say that you "can't tell" about student understanding because there is no evidence that would help you with diagnosis. Previous research has shown this is a hard skill for PTs because they tend to be overly generous when evaluating student understanding (see, e.g., Morris, 2006). However, in the messy environment of the classroom, it is necessary to be able to recognize evidence from non-evidence in order to accurately diagnose student understanding.

Also following Spitzer et al. (2011), this was a quasi-experimental design with no control group. This was due to the practical and ethical constraints of the classroom, where PTs could not be randomly assigned to either an intervention that might improve their overall course grade or a control group that would not. In fact, most intervention work on PTs' diagnostic competence to-date has been quasi-experimental. This means further research is needed, including replication studies

Table 1 Timeline of data collection and intervention

Time	Participant activities
Week 3, homework, approximately 1 hour	Pretest: Complete individual transcript analysis task
Week 4, in class, approximately 30 minutes	Intervention: Individual work on new transcript activity
Week 4, in class, approximately 30 minutes	Intervention: Group work on new transcript activity
Week 4, in class, approximately 30 minutes	Intervention: Whole class discussion on new transcript activity with main ideas made clear by instructors
Week 4, in class, approximately 45 minutes	Posttest: Complete transcript analysis task in groups

such as this one. In quasi-experimental design, it is important to address threats to internal validity; we have done so below.

One difference in procedure from Spitzer et al. (2011) was that due to the constraints of a teaching experiment, where data collection must occur in the midst of a lesson, the posttest was administered in groups of 2 or 3 students. That is, students completed the pretest individually and the posttest in groups. This makes the posttest a more supportive task than the pretest. To account for this and allow for comparison between the pre- and posttest, we did not compare individual scores and instead compared group scores on the pre- and posttest. That is, we grouped PTs' pretest scores into the same groups they were in for the posttest. We then compared the mean group scores on the pretest to the group scores on the posttest. Thus, the unit of analysis (for both pre- and posttest) was a group of 2 or 3 PTs ($N = 10$ groups).

4.3 Intervention

The timeline of the data collection and intervention activities is presented in Table 1. Previous research shows that PTs do a better job diagnosing and evaluating student thinking with familiar mathematics (Bartell et al., 2013; Phelps & Spitzer, 2012). Thus, the intervention focused on mathematics which PTs knew well and the class included a mathematics review activity prior to the intervention.

The intervention occurred in class during week 4 of the semester and consisted of PTs working first individually, then in groups, then as a whole class on analyzing a lesson on comparing fractions. This lesson was different from the pre- and posttests but used the same learning goal to avoid introducing new mathematics. As PTs analyzed the lesson as a whole class, they also had some of the main ideas made clear by the instructor. In particular, the instructor emphasized that

- Procedural work is not evidence we can use to diagnose students' conceptual understanding.

- Correct answers are not sufficient evidence to diagnose full understanding of the learning goal.
- While important, a teacher explaining the main mathematical ideas provides no evidence for diagnosing student understanding.
- Only student work aligned with the learning goal can help us diagnose their understanding of the learning goal.

In this conceptual replication study, we aim to build on and add to the theory of the previous research (Makel & Plucker, 2014). Thus, we made several changes to the intervention based on Spitzer et al.'s (2011) results. First, in contrast with Spitzer et al. (2011), the intervention we present here did not include a card sort task in which PTs were asked to align evidence with learning goals. Follow-up research suggests this card sort task was a supportive context for PTs' analyzing evidence of student thinking (Phelps & Spitzer, 2012). However, we wondered if this more supportive task might be less effective in allowing PTs to transfer their skills to the messy and confusing environment of a classroom. In addition, time in teacher education is a valued commodity. Thus, we were interested to see if a shorter intervention, which did not feature the supportive card sort task, might still be effective in improving PTs' ability to diagnose evidence of student thinking. Thus, we focused solely on a transcript task which included some distractors and evidence that was non-revealing of student thinking.

Several additional changes to Spitzer et al.'s (2011) work were made. In particular, we did not include a "minimal intervention," where PTs were told that the lesson was unsuccessful and asked to reconsider their ideas. Phelps and Spitzer (2012) found that such an intervention was not beneficial in improving PTs' diagnostic and evaluation skills. In addition, while we included group work throughout the intervention, we did not ask students to discuss their journal assignment in groups until after the intervention, when they had more of an explicit framework for evaluating their work. Previous research found that asking students to engage in group work prior to any whole class discussion was not beneficial in improving PTs' diagnosis of the evidence (Phelps & Spitzer, 2012).

5 Results

We first explore convergent results and then explore divergent results compared to Spitzer et al. (2011); both show how our findings support and extend the work of Spitzer et al. (2011).

5.1 Convergent Results

Analysis of the pre- and posttests indicates that PTs did improve in their ability to diagnose evidence of student thinking in a classroom lesson. As noted above, PTs' responses were coded as a 0, 1, or 2, with 0 being the best response for all six sections of the pre- and posttests. For each group of PTs, we compared the group score

Table 2 Mean ratings for the pre- and posttests

Excerpt	Pretest mean (s)	Posttest mean (s)	Mean difference (std. error mean)	Significance level
Teacher explanation	0.74 (0.19)	0.50 (0.00)	0.24 (0.06)	$p < 0.01$
Irrelevant – picture	0.93 (0.52)	0.50 (0.85)	0.43 (0.28)	$p > 0.05$
Irrelevant – calculator	0.88 (0.34)	0.00 (0.00)	0.88 (0.11)	$p < 0.001$
Procedural – detailed	1.92 (0.18)	1.2 (0.42)	0.72 (0.13)	$p < 0.001$
Procedural – not detailed	0.90 (0.32)	0.40 (0.52)	0.50 (0.16)	$p < 0.05$

on the pretest to the group score on the posttest. Results indicate that, looking across the whole transcript, PTs' scores improved from the pretest to the posttest. Specifically, the average rating for each section of the transcript decreased significantly after the intervention, from 1.02 ($s = 0.16$) to 0.52 ($s = 0.21$) ($p < 0.001$). This indicates that overall, groups of PTs changed from an average response of "can tell a little" to about halfway between "can't tell" and "can tell a little." Since the best answer for each section is "can't tell" about student achievement of the learning goal, this is a positive result, and is in line with the results of Spitzer et al. (2011).

Having found that the intervention appeared to improve PTs' performance across the entire transcript, we were interested to see how their performance changed for each of the three types of evidence in the lesson sample: Teacher explanations, procedural descriptions, and irrelevant responses. Table 2 provides a summary of the mean ratings for each section of the pre- and posttests, and the results for each type of evidence are discussed below.

5.1.1 Evidence Consisting of Teacher Explanations

In the pre- and posttest transcript, two sections consisted solely or mostly of the teacher explaining the mathematics followed by a lack of student questions or students saying "Oh, I see." Although teacher explanations might help students learn, they do not provide any evidence that students have actually achieved a learning goal. Furthermore, students may say, "Oh, I see" or refrain from asking questions for many reasons, not necessarily because they have a full understanding of the ideas. As Hiebert et al. (2007) note, "It is tempting to assess teaching effectiveness based on what the teacher does rather than on how the students respond" (p. 52), but competent diagnoses of student thinking require that teachers examine student responses. Thus, the best response for these sections was "I can't tell about student understanding."

We found that PTs were initially fairly positive about teacher explanations on the pretest, but after the intervention, they were less likely to judge a teacher explanation as evidence of student thinking. As Table 2 indicates, the mean rating for this type of evidence decreased significantly from 0.75 on the pretest to 0.5 ($p < 0.01$). This result is consistent with the findings of Spitzer et al. (2011).

5.1.2 Evidence Containing Strictly Procedural Descriptions

Our pre- and posttests each contained two sections where students correctly solved a fraction comparison problem using common denominators. However, their solutions failed to address the learning goal because they were strictly procedural and showed no understanding of the key concepts underlying the strategy (as the learning goal specified). In one of these sections, the comparison problem already had common denominators (compare ¼ and ¾). The student in the transcript section correctly solved this problem by analyzing the numerator; however, his thinking in doing so is unclear because the section lacks detail.

Consistent with Spitzer et al. (2011), the intervention appears to have improved PTs' ability to diagnose evidence of student thinking on this non-detailed procedural section. On the pretest, the mean score on this section was 0.90, with 2 of the 22 (9%) PTs ranking the section as "can't tell," the best response. On the posttest, the mean score on this section was 0.4, with 6 of the 10 (60%) groups of PTs ranking this section as "can't tell," the best response. This significant change suggests that the intervention was successful in helping PTs look beyond the mere presence of a correct answer when diagnosing evidence of student thinking.

5.2 *Divergent Results*

The intervention improved PTs' abilities in two areas Spitzer et al. (2011) found no statistically significant improvement. Additionally, the intervention failed to improve PTs' abilities in one section where Spitzer et al. (2011) saw statistically significant improvement. Each of these instances of divergence is discussed below.

5.2.1 Evidence Irrelevant to the Learning Goal

Following Spitzer et al. (2011), the pre- and posttest transcript included two sections of evidence that was irrelevant to the learning goal. In one section, the student correctly solved a fraction comparison problem by using a calculator to convert to decimals. In the other, the student correctly solved a fraction comparison problem by using a pie-chart picture representation. Both of these sections contain students' correct answers, but no evidence of student thinking that aligns with the ideas of the learning goal (the concepts of the common denominator strategy).

For the section in which a student uses a calculator to correctly solve the fraction comparison problem, the intervention significantly improved PTs' ability to diagnose evidence of student thinking. On the pretest, the mean for this section was 0.88 ($s = 0.34$) and on the posttest it was 0.00 ($s = 0.00$) ($p < 0.001$). This means that, on the posttest, every group of PTs ranked the section as "can't tell." This is a divergent result from Spitzer et al. (2011) who found that the decrease in scores was not significant for this section. Spitzer et al. (2011) suggest their scores did not significantly decrease because PTs developed an anti-calculator sentiment, where they assumed that use of a calculator told you strongly ("can tell a lot") that students did not understand. This divergent result is important because it suggests that PTs in the current intervention did not develop this strong anti-calculator sentiment.

For the section consisting of a student using a picture to correctly solve the problem, the intervention does not appear to have significantly improved PTs' ability to diagnose evidence of student thinking. On the pretest, the mean rating for this section was 0.93 and on the posttest the mean rating was 0.5 ($p > 0.05$). This is a divergent result from Spitzer et al. (2011), whose intervention did significantly improve PTs' ability to analyze a student solving this problem with a picture. However, this might be a divergent result caused by lack of statistical power. Specifically, Spitzer et al.'s (2011) sample size was bigger, potentially allowing them to find significant results from smaller gains.

5.2.2 Evidence Containing Strictly Procedural Descriptions

Recall that our pre- and posttests contained two sections where students correctly solved a fraction comparison problem using common denominators but in a strictly procedural way. The first of these sections (procedures lacking detail) was discussed above. In the second procedural evidence section, the student correctly solves the problem and provides a detailed account of his procedural steps but without any reference to the concepts underlying those steps, and thus with no evidence of understanding.

On the procedural, detailed section, the intervention improved PTs' abilities, with the mean decreasing significantly from 1.92 ($s = 0.18$) to 1.2 ($s = 0.42$) ($p < 0.001$). As in the Spitzer et al. (2011) study, PTs were initially very positive about student achievement for this section. On the pretest, no PT ranked the section as "can't tell," three PTs (14%) ranked it as "can tell a little," and the remaining 19 PTs (86%) ranked it as "can tell a lot." This suggests that, on the pretest, PTs were very willing to attribute conceptual understanding where they had evidence to diagnose only procedural competency. In contrast, on the posttest, 8 of the 10 groups (80%) ranked the section as "can tell a little" and only 2 of the groups ranked the section as "can tell a lot." This is a divergent result from Spitzer et al. (2011), who did not find a statistically significant decrease in PTs' scores after the intervention. However, both the original and replication study found that PTs continued to be fairly positive about procedural evidence even after an intervention, ranking this section as most revealing in the whole transcript.

6 Discussion

Our primary goal for the present study was to replicate the work of Spitzer et al. (2011). A key hypothesis of both the current and previous study (Spitzer et al., 2011) is that teacher educators can use interventions to improve one aspect of diagnostic competence, specifically PTs' ability to distinguish evidence that can be used to diagnose students' understanding of a mathematical learning goal from non-evidence. Both this and the previous study confirm this hypothesis, showing teacher educators can use interventions to significantly improve PTs' ability to diagnose evidence of student thinking.

6.1 Limitations

Our study design followed Spitzer et al. (2011) closely because it was a conceptual replication. One design choice of Spitzer et al. (2011) was to have a quasi-experimental pre- and posttest design with no control group. This choice was necessary for practical and ethical reasons in both the original and replication studies. In addition, the advantage of quasi-experimental work is that the results are more readily generalized to applied settings such as teacher education programs (Reichardt, 2009). However, it is important to address the limitations of this type of design, particularly potential threats to internal validity.

Reichardt (2009) identifies seven threats to internal validity that researchers must consider in this type of design. However, as Reichardt (2009) notes, "none of the threats to internal validity may be plausible in studies of educational interventions that teach materials that are unlikely to be learned elsewhere, where the pretest and posttest measures focus solely on the material being taught, [and] where the time interval between pretest and posttest is short" (p. 50). We considered these threats in our design and do not believe they are applicable here. In particular, since diagnostic competence is not regularly taught in teacher education in the U.S., PTs are unlikely to have learned it elsewhere between the pre- and posttest.

6.2 Conceptual Replication

Our study was a conceptual replication of Spitzer et al. (2011), meaning it differed in some methodological ways from the original study with the goal of extending the underlying ideas and generalizability of the original study. Thus, for example, the PTs in this study were mathematics majors and minors, in contrast with Spitzer et al.'s (2011) participants, who were general elementary education students with weaker mathematics backgrounds. It is thus reasonable to assume that the participants in the replication study had stronger mathematics backgrounds and potentially more robust mathematics knowledge for teaching (see Ball, Thames, & Phelps, 2008). This is important because previous work has shown that diagnostic competence appears to be related to mathematics knowledge for teaching (e.g., Bartell et al., 2013; Philipp & Leuders, 2014; Schack et al., 2013). In spite of this, the PTs' skills in this replication still improved post-intervention. This builds on the results of Spitzer et al. (2011) and indicates that interventions can be used to improve PTs' diagnostic abilities regardless of their year in university or their mathematical background.

Hiebert et al. (2007) argue that evidence used to diagnose student achievement should be both relevant to the learning goal and revealing of students' understanding. Spitzer et al. (2011) found that their intervention improved PTs' ability to recognize relevant evidence but did not improve their ability to recognize revealing evidence, particularly to distinguish detailed procedural work from conceptual understanding. In contrast, we found the intervention improved PTs' abilities on both types of evidence. While this may be due to some subtle change in the intervention, we instead hypothesize that this is the result of the difference in populations. In particular, the PTs in the present study probably had stronger mathematical content knowledge which may have led to these PTs being more able to distinguish detailed procedural evidence and evidence of conceptual understanding. Future intervention work could focus on the difference mathematics knowledge makes on PTs' ability to diagnose evidence of student thinking.

This work, which focuses on a certain aspect of diagnostic competence, adds to previous work, which shows that PTs' ability to diagnose student understanding can be improved through intervention (e.g., Alsawaie & Alghazo, 2010; Santagata & Angelici, 2010; Santagata & Guarino, 2011; Santagata et al. 2007; Spitzer & Phelps, 2011; Stockero, 2008). The intervention of the present study included several key features common to interventions that have been found to improve PTs' abilities. First, it used a representation of practice, in this case a transcript of a classroom lesson. Second, while PTs were not given a written framework for analysis, they engaged in a discussion which explicitly included attention to guidelines such as procedural work is not evidence of conceptual understanding and correct answers are not sufficient evidence for diagnosing student understanding. The inclusion of the teacher educator providing expert ideas is important because previous work has shown PTs need to have the main ideas made explicit by an expert (Phelps & Spitzer, 2012). This also aligns with general recommendations that students must grapple with conceptual ideas but also have those ideas made explicit by instructors (Hiebert & Grouws, 2007). Finally, the intervention was of a short length (approximately one class week). In fact, the intervention of the present study was slightly shorter than that of the Spitzer et al. (2011) study, including slightly less class time and not including a card sort or "minimal" intervention (where PTs were told the lesson was unsuccessful and asked to revisit their work).

In line with Spitzer et al. (2011), we found that a well-designed intervention can improve PTs' ability to correctly appraise diagnostic evidence of student thinking, regardless of the PTs' prior experiences or mathematics background. In fact, the intervention in the current study improved PTs' diagnostic abilities while being shorter, an important result because it suggests this is a teachable skill even in busy teacher education programs. While not all teacher education programs may have room for a semester-length intervention like the one used by Santagata and Yeh (2014), the majority of programs should have room for short interventions which may result in significant improvement in PTs' diagnostic abilities.

Although the research base now seems to firmly indicate that interventions have the potential to improve PTs' diagnostic abilities, some open questions remain. For example, future research could examine the differing impacts of shorter and longer interventions for improving PTs' abilities, or use several comparison groups to more precisely pinpoint the kinds of experiences which most effectively improve PTs' skills in diagnosing student thinking. Finally, more research is needed to investigate the ways in which PTs can bring these skills into their classroom and use them to truly become lifelong learners who continuously improve their practice over time.

References

Adler, J., Ball, D., Krainer, K., Lin, F. L., & Novotna, J. (2005). Reflections on an emerging field: Researching mathematics teacher education. *Educational Studies in Mathematics, 60*(3), 359–381.

Alsawaie, O. N., & Alghazo, I. M. (2010). The effect of video-based approach on prospective teachers' ability to analyze mathematics teaching. *Journal of Mathematics Teacher Education, 13*(3), 223–241.

Ball, D. L., Thames, M. H., & Phelps, G. (2008). Content knowledge for teaching: What makes it special? *Journal of Teacher Education, 59*(5), 389–407.

Bartell, T. G., Webel, C., Bowen, B., & Dyson, N. (2013). Prospective teacher learning: Recognizing evidence of conceptual understanding. *Journal of Mathematics Teacher Education, 16*(1), 57–79.

Berk, D., & Hiebert, J. (2009). Improving the mathematics preparation of elementary teachers, one lesson at a time. *Teachers and Teaching: Theory and Practice, 15*(3), 337–356.

Feiman-Nemser, S. (2001). From preparation to practice: Designing a continuum to strengthen and sustain teaching. *Teachers College Record, 103*(6), 1013–1055.

Hammerness, K., Darling-Hammond, L., & Bransford, J. (with Berliner, D., Cochran-Smith, M., McDonald, M., & Zeichner, K.) (2005). How teachers learn and develop. In L. Darling-Hammond & J. Bransford (Eds.), *Preparing teachers for a changing world: What teachers should learn and be able to do* (pp. 358–389). San Francisco, CA: Jossey-Bass.

Hiebert, J., & Grouws, D. A. (2007). The effects of classroom mathematics teaching on students' learning. In F. K. Lester (Ed.), *Second handbook of research on mathematics teaching and learning* (pp. 371–404). Charlotte, NC: Information Age Publishing.

Hiebert, J., Morris, A. K., Berk, D., & Jansen, A. (2007). Preparing teachers to learn from teaching. *Journal of Teacher Education, 58*(1), 47–60.

Jacobs, V. R., Lamb, L. L. C., & Philipp, R. A. (2010). Professional noticing of children's mathematical thinking. *Journal for Research in Mathematics Education, 41*(2), 169–202.

Leuders, T., Dörfler, T., Leuders, J., & Philipp, K. (2018). Diagnostic competence of mathematics teachers – unpacking a complex construct. In T. Leuders, K. Philipp, & J. Leuders (Eds.), Diagnostic competence of mathematics teachers. Cham: Springer.

Makel, M. C., & Plucker, J. A. (2014). Facts are more important than novelty: Replication in the education sciences. *Educational Researcher, 43*, 304–316.

Meikle, E. (2014). Preservice teachers' competencies to select and sequence students' solution strategies for productive whole-class discussions. *Mathematics Teacher Educator, 3*(1), 27–57.

Morris, A. K. (2006). Assessing pre-service teachers' skills for analyzing teaching. *Journal of Mathematics Teacher Education, 9*, 471–505.

Morris, A. K., Hiebert, J., & Spitzer, S. M. (2009). Mathematical knowledge for teaching in planning and evaluating instruction: What can preservice teachers learn? *Journal for Research in Mathematics Education, 40*(5), 491–529.

Phelps, C. M., & Spitzer, S. M. (2012). Systematically improving lessons in teacher education: What's good for prospective teachers is good for teacher educators. *The Teacher Educator, 47*(4), 328–347.

Philipp, K., & Leuders, T. (2014). Diagnostic competences of mathematics teachers – Processes and resources. In *Proceedings of the 38th Conference of the International Group for the Psychology of Mathematics Education, Vol. 4*, Vancouver, BC, Canada.

Reichardt, C. S. (2009). Quasi-experimental design. In R. E. Millsap & A. Maydeu-Olivares (Eds.), *The SAGE handbook of quantitative methods in psychology* (pp. 46–71). Los Angeles, CA: Sage Publications.

Santagata, R., & Angelici, G. (2010). Studying the impact of the lesson analysis framework on preservice teachers' abilities to reflect on videos of classroom teaching. *Journal of Teacher Education, 61*(4), 339–349.

Santagata, R., & Guarino, J. (2011). Using video to teach future teachers to learn from teaching. *ZDM, 43*(1), 133–145.

Santagata, R., & Yeh, C. (2014). Learning to teach mathematics and to analyze teaching effectiveness: Evidence from a video- and practice-based approach. *Journal of Mathematics Teacher Education, 17*(6), 491–514.

Santagata, R., Zannoni, C., & Stigler, J. W. (2007). The role of lesson analysis in pre-service teacher education: An empirical investigation of teacher learning from a virtual video-based field experience. *Journal of Mathematics Teacher Education, 10*, 123–140.

Schack, E. O., Fisher, M. H., Thomas, J. N., Eisenhardt, S., Tassell, J., & Yoder, M. (2013). Prospective elementary school teachers' professional noticing of children's early numeracy. *Journal of Mathematics Teacher Education, 16*, 379–397.

Schmidt, S. (2009). Shall we really do it again? The powerful concept of replication is neglected in the social sciences. *Review of General Psychology, 13*, 90–100.

Schoenfeld, A. H. (2011). Noticing matters. A lot. Now what? In M. G. Sherin, V. R. Jacobs, & R. A. Philipp (Eds.), *Mathematics teacher noticing: Seeing through teachers' eyes* (pp. 223–238). New York, NY: Routledge.

Sherin, M. G., Jacobs, V. R., & Philipp, R. A. (2011). Situating the study of teacher noticing. In M. G. Sherin, V. R. Jacobs, & R. A. Philipp (Eds.), *Mathematics teacher noticing: Seeing through teachers' eyes* (pp. 3–13). New York, NY: Routledge.

Sherin, M. G., Linsenmeier, K. A., & van Es, E. A. (2009). Selecting video clips to promote mathematics teachers' discussion of student thinking. *Journal of Teacher Education, 60*, 213–230.

Spitzer, S. M., & Phelps, C. M. (2011). Prospective teachers' use of lesson experiments: The effects of an online discussion board. In L. R. Wiest & T. Lamberg (Eds.), *Proceedings of the 33rd annual meeting of the North American Chapter of the International Group for the Psychology of Mathematics Education*, Reno, NV.

Spitzer, S. M., Phelps, C. M., Beyers, J. E. R., Johnson, D. Y., & Sieminski, E. S. (2011). Developing pre-service elementary teachers' abilities to identify evidence of student mathematical achievement. *Journal of Mathematics Teacher Education, 14*(1), 67–87.

Star, J., & Strickland, S. (2008). Learning to observe: Using video to improve pre-service mathematics teachers' ability to notice. *Journal of Mathematics Teacher Education, 11*, 107–125.

Stockero, S. (2008). Differences in preservice mathematics teachers' reflective abilities attributable to use of a video case curriculum. *Journal of Technology and Teacher Education, 16*(4), 433–458.

Specific Mathematics Assessments that Reveal Thinking: An Online Tool to Build Teachers' Diagnostic Competence and Support Teaching

Kaye Stacey, Vicki Steinle, Beth Price, and Eugene Gvozdenko

In this chapter, we describe the design of an online system for the formative assessment of students' understanding of mathematics and discuss how it develops diagnostic competence and influences teaching. The smart-test system covers many mathematics topics studied by students between about 10 and 16 years of age. It is programmed to provide teachers with an automated diagnosis of their own students' stages of development in specific topics and to report on an individual's errors and misconceptions, in order to inform teaching. Our claim is that teachers' diagnostic competence increases when they have easy access to information about their own students' thinking. In turn, this can further improve teaching, and hence learning. By drawing together evaluative data from four sources, we highlight aspects of teachers' initial responses to formative assessment and the effect of using this system on their knowledge for teaching and the subsequent changes to teaching practice. Overall, teachers report that using the smart-tests has improved their knowledge of the thinking of individual students as well as of students in general (i.e., their pedagogical content knowledge), and that they can use this information in several ways to adjust their teaching. Paradoxically, using smart-tests reduces the demand for teachers to have specific knowledge for diagnosis and at the same time increases this knowledge and so improves their diagnostic competence.

1 Introduction

This chapter discusses the design and use of an online assessment system and presents a partial evaluation of the success of this system. The smart-test system is built on *Specific Mathematics Assessments that Reveal Thinking,* which we abbreviate as

K. Stacey (✉) • V. Steinle • B. Price • E. Gvozdenko
University of Melbourne, Melbourne, VIC, Australia
e-mail: k.stacey@unimelb.edu.au

T. Leuders et al. (eds.), *Diagnostic Competence of Mathematics Teachers*, Mathematics Teacher Education 11, DOI 10.1007/978-3-319-66327-2_13

smart-tests. These tests are accessed through an intelligent environment (HREF1), created by the authors (Stacey, Price, Steinle, Chick, & Gvozdenko, 2009). The goal is to diagnose individual student's understanding of mathematics topics, hence improve the teacher's understanding of student thinking, and thereby assist teachers to target lessons to better meet the needs of their students.

The smart-test system provides teachers with an informative diagnosis of their students' conceptual understanding of many of the topics in the curriculum for students between 10 and 16 years of age. The diagnoses are described in terms of developmental stages and the misconceptions and/or common errors that have been identified for a particular student. These diagnoses are available to a teacher immediately after their students complete the test. As far as possible, the items, the developmental frameworks, and the diagnostics are drawn from well-established research findings and so build in pedagogical content knowledge related to student thinking. In addition, the system provides teachers with explanations of the diagnoses, teaching suggestions for taking students to the next level of understanding, and, if appropriate, for dealing with misconceptions and common errors. Although the categories are not well defined, we find it useful to distinguish between misconceptions (which have an identifiable underlying conceptual base) and common errors (e.g., information that students have not learned, confusions of names, and bugs in algorithms) that are procedurally or factually based.

Because smart-tests aim to assist the teacher to plan more effective teaching, a smart-test is focused on one topic and typically takes students less than 10 min to complete. To encourage teachers to monitor student progress throughout the teaching of a topic, there are two parallel test versions for each topic. Smart-tests are not intended to be a complete assessment of the topic – for example, there are no lengthy items which require students to write mathematical reasoning. The smart-test system is currently being used regularly by over 400 teachers, and we process approximately 7000 student tests each month. The system can be used by teachers and students anywhere with an internet connection.

The smart-test system is designed to assist teachers with the diagnostic activities of gathering and interpreting data so that they obtain valid knowledge on the achievement of individual students and to provide appropriate teaching suggestions as a basis for action. As will be evident in the sections below, making use of this information involves the whole range of diagnostic competence, including teachers' knowledge (especially their pedagogical content knowledge), beliefs connected to formative assessment and the skills to implement it and to act on the findings. We will demonstrate how the smart-test system paradoxically reduces the demand for teachers' diagnostic competence, whilst at the same time building it.

Section 2 outlines the designers' vision. In creating any educational product, there is a myriad of design decisions, so this paper concentrates on those that are central to providing diagnostic judgments that can support productive action by teachers. Sections 3 and 4 draw together feedback from teachers gathered from several different sources over the life of the project. In Sect. 3, we discuss two themes related to teachers' evolving understanding of the use of formative assessment – what it is really for and how it is best used and discuss how we responded to these issues. In Sect. 4, we report on our progress towards achieving the two fundamental goals of the smart-test system, namely higher achievement for students through targeted teaching, and improved mathematical pedagogical content knowledge for teachers.

1.1 Data Sources

This chapter reports experiences of the smart-test creators and feedback from users. The data reported in Sects. 3 and 4 is a collation from four sources: (1) records of 10 focus groups held with teachers at three schools involved in the development of the smart-test system in its first 2 years (2008–2009); (2) online surveys completed by volunteer teachers after they have used a smart-test (2009–2014); (3) spontaneous emails that teachers have sent to us on an ad-hoc basis after completing a smart-test (2009–2015); and (4) interviews with three mathematics leaders. Two of the interviewees were teachers holding leadership positions in mathematics at their schools (Leader 1 and Leader 2); they were interviewed in 2015. The third interviewee was a Project Officer, employed by an education authority, who assisted teachers using the smart-tests in their own schools as part of a larger professional development learning programme. She was interviewed in both 2012 and 2015.

Gaining feedback on the smart-test system is in itself a process of formative evaluation, with the aim of improving all aspects of the system. As is evident from the sources described above, the data sources reflect the long development time for this complex resource. Because of ongoing improvements, the resource to which they responded is somewhat different at each stage with early concerns having now been addressed. The feedback reported in Sects. 3 and 4 focusses on issues that transcend pragmatic concerns (e.g., difficulty scrolling on long pages, download speeds) and gets to the heart of how teachers might use formative assessment from any source to improve their teaching.

2 The Designers' Vision for the Smart-Test System

> *…because learning is unpredictable, assessment is necessary to make adaptive adjustments to instruction.* (Wiliam, 2011, p. 13)

The initial concept of the smart-test system arose from our observation of the power of formative assessment and our observation of the difficulty of diagnosing students' thinking quickly and efficiently. This is well supported by others (e.g., Wiliam, 2007, 2011). In previous research projects, we saw how teaching about decimal numeration could be transformed by giving teachers information about the ways in which each of their students thought about decimal notation and by helping teachers understand the common misconceptions (Helme & Stacey, 2000). However, we also noted that it is time-consuming for teachers to diagnose individual student thinking using research level tests (Steinle, 2004) or simplified versions (Steinle, Stacey, & Chambers, 2006). Diagnosis from written tests is usually complicated because special scoring instructions must be followed to identify the telltale patterns in students' responses that indicate a misconception. Teachers expect to mark students' work as correct or incorrect, and to make total scores or sub-scores, but it is beyond their expectations that they should undertake further processing of data, such as examining patterns of responses rather than just observing direct errors in a systematic way. The solution to this dilemma was to use online assessment, with computer programming identifying the patterns of responses across multiple items that reveal thinking. Hence, the smart-test system began.

The need for technological help in formative assessment has been noted by others. For example, Pellegrino and Quellmalz (2010) wrote:

> No individual, whether a classroom teacher or other user of assessment data, could realistically be expected to handle the information flow, analysis demands, and decision-making burdens involved without technological support. Thus, technology removes some of the constraints that previously made high-quality formative assessment difficult or impractical for a classroom teacher. (p. 130)

We planned that this system would be easy and efficient for teachers to use and that it would supply information that is concise enough to be readily useable by teachers, sufficiently valid and deep enough to make a real difference to lesson content, and linked to targeted teaching resources. Figure 1 shows how we expected teachers to interact with the system and the two predicted outcomes: higher achievement for students through targeted teaching, and improved mathematical pedagogical content knowledge for teachers. We expected that this aspect of knowledge for teaching would improve as teachers become familiar with the developmental stages and possible misconceptions in a particular topic, especially as they see how these apply to their own students.

The smart-test system embeds research in mathematics education into artefacts that are intended to be easy for practitioners to use, creating what Pea (1987) calls 'distributed intelligence' in tools for teaching. When planning the teaching of a new topic, the diagnostics from the system provide teachers with knowledge of the mathematical thinking of their current students. It is intended that, simultaneously, teachers will also learn about the mathematical knowledge of students more generally and hence will be better able to teach effectively in the future.

There are considerable benefits if a teacher is able to conduct interviews with students on their mathematical understanding. Indeed, interviewing all students to establish their stages of development has been a central feature of the highly effective early numeracy programmes in Australia, such as Count Me In Too (Stewart, Wright & Gould, 1998). There are further examples in the chapter by Clarke et al. in this book. However, interviewing is a resource-intensive option, for which schools need to make very special arrangements. We make no claim that smart-test information is always completely accurate, but neither is any other method (although this is sometimes not recognised!) and teachers can choose to talk to those few students

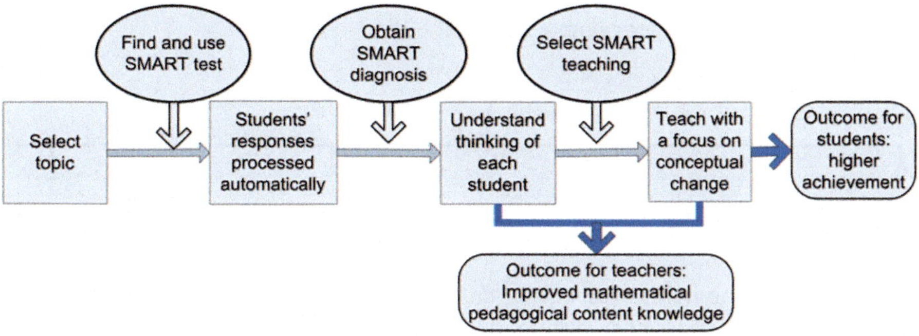

Fig. 1 Using the smart-test system and its predicted outcomes

with puzzling smart-tests results. The intention is that the smart-test system should provide teachers with sufficiently valid information to influence the teaching of topics about to be taught, in a timely and simple manner.

2.1 *Items that Provide a Window into Student Thinking*

> *Items that reveal unintended conceptions – in other words that provide a 'window into thinking' – are not easy to generate, but they are crucially important to improve the quality of students' mathematical learning.* (Wiliam, 2007, p. 1069)

Smart-tests are built on exactly the sorts of items that Wiliam refers to in the quote above: sets of items that together provide a window into student thinking. Over the several decades of mathematics education research into students' thinking, a rich bank of items has been established which smart-tests make more accessible to classroom practice. However, items usually need to be modified for use in computer-based assessment because a computer is still limited in its processing of free response items (Stacey & Wiliam, 2013). Hence smart-tests often include selected response items with alternatives based on research evidence.

In addition to the multiple choice format, there are now other selected response formats that can be readily computer marked. Sliders provide a very flexible interactive format. Students can place numbers on number lines as in Fig. 2, and show estimates of quantities like percentages and angles. Drag-and-drop items allow students a different type of participation, similar to the way they might participate in an interview. For example, students can arrange 'cards' showing various fractions and decimals in ascending order; they can place a card showing the position of the translated image of an object; and they can drag cards showing representations of the fraction two-thirds into one pile, and other cards into another. Figure 3 shows a student's incorrect pattern of responses to an item about sorting angles. Using this expanded range of computer-assessable formats has made the tests more interactive.

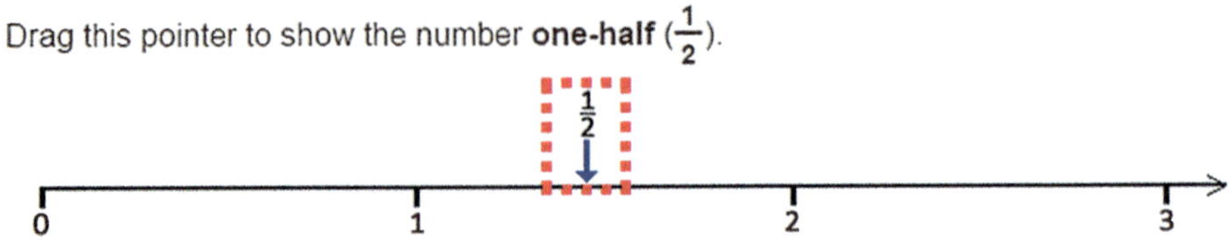

Fig. 2 A student's incorrect response to a slider item

As well as using items with strong credentials from the research literature, sets of items for smart-tests must systematically vary the features which are known to make a difference to item success rates. This enables students' difficulties to be pinpointed. It also gives guidance to teachers about the range of items to include in instruction and what constitutes robust understanding. Using only prototypical item types in teaching is known to encourage misconceptions and limited understanding. For example, when considering angles of the same size, factors such as the ray length shown and the orientation of the angles affect success rates. The drag-and-drop item described earlier identifies whether students see the angle (the amount of turn) by systematically varying distracting factors. Figure 3 shows the display seen by a teacher for one of their own student's choices of angles of the 'same size'. The borders that can be seen in this figure are to assist teachers when they look at test submissions; green is correct and red is incorrect. Note that this student, like many others, has incorrectly used the visible ray length to classify the angles. This student has done this consistently; others may nearly always do it. Teachers are provided with a brief summary of findings for each student, and have the option of accessing student screens to see the behaviour in action.

2.2 *Diagnosis from Patterns of Responses, Not Just Score*

Many mathematics tests base their assessment only on accuracy, either total test score or scores on subsections. The power of the smart-test system is that it diagnoses student thinking based on the actual responses. Responses (more than accuracy) are used because 'wrong in which way' is more revealing than just 'wrong'. For example, the student whose work is shown in Fig. 3 might be considered to have a score of 3 out of 9, but the pattern of wrong answers actually shows *why* they were wrong, not just that they *were* wrong. As far as possible, smart-tests report on the reasons for errors, not just the number of items correct.

Steinle, Gvozdenko, Price, Stacey, and Pierce (2009) indicate how response patterns can be used to diagnose student misconceptions in algebra in the test named *Values for letters*, which draws on extensive research in algebra such as Küchemann (1981) and Fujii (2003). Those who sign up on the smart-test website can access the 2012 version of this test which replaced the 2009 version. One set of items in the test *Values for letters* describes the scenario that 'some students' were asked to find

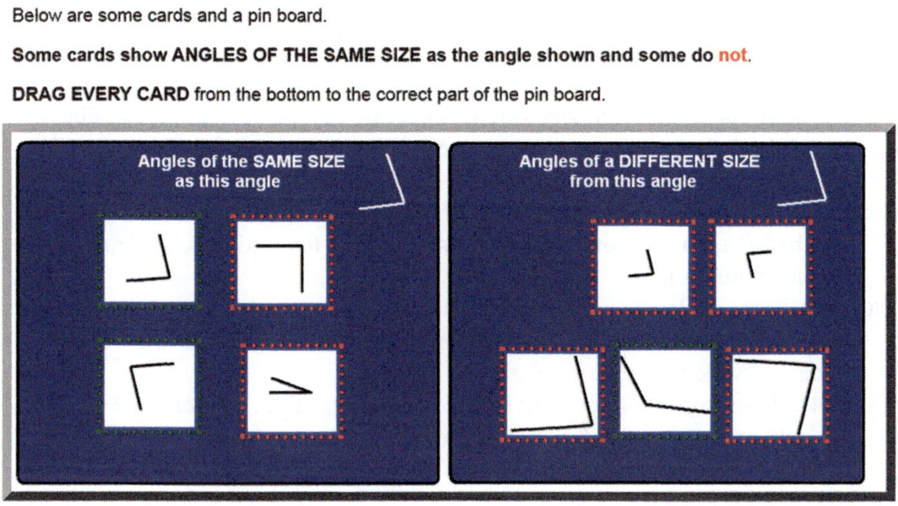

Fig. 3 A student's incorrect response to understanding angle size item

the values of letters in several equations. In the first item, the students taking the test are asked to indicate whether the solutions to $x + x + x = 12$ given by the fictional students are right or wrong. For example, one fictional student has answered '$x = 2$ and $x = 5$ and $x = 5$' (which is incorrect but accepted by students who see the letter as simply a placeholder for any number) and another has answered '$x = 4$' (which is actually correct, but will be rejected by students who want a value for each of the three occurrences of x). In the second item, the equation is $x + y = 16$ and the solutions given by the fictional students include '$x = 7$ and $y = 9$' which is correct, although rejected by the few students who think y should be one more than x because y is one letter after x in the alphabet (Stacey & MacGregor, 1997) and '$x = 8$ and $y = 8$' (also correct but rejected by students who believe that different letters must stand for different numbers). Based on the pattern of their responses, students are allocated to one of four developmental stages (see Fig. 4) for interpreting the letters in equations and are flagged when they have certain misconceptions. Smart-test items are often, as in this case, drawn from research literature, but using the tests also provides data on the prevalence of reported misconceptions in our population. In a few cases, previously unreported misconceptions have been revealed.

The only situation where we use the score (total number of responses correct) to provide information to teachers is for an additional feature introduced in response to feedback from teachers about anomalous results. Occasionally a student fails to clear the hurdles for the early stages, yet answers more difficult items correctly. Students with a high score on a test but a low developmental stage are therefore flagged so that teachers can investigate individually whether the students do have a fundamental misunderstanding or gap in their knowledge, perhaps masked by procedural expertise, or have just made some careless errors or omissions. We have observed examples of both situations. As an extreme example, a student who did not

Developmental stages for smart-test *Values for letters*

Stage 1: The students know that letters can stand for numbers, and are able to correctly substitute into very simple algebraic expressions, but they believe that the values that letters can take are in some way related to their place in the alphabet.

Stage 2: These students interpret an algebraic letter only as a place holder for a number in a number sentence, so they allow one letter to have several values in one expression.

Stage 3: These students appreciate that each time a particular letter is used in an equation it stands for the same number, but they over-generalise to 'different letters must be different numbers'.

Stage 4: These students know that in one algebra question, a letter must stand for only one number and that different letters can stand for the same number.

Misconceptions and common errors for *Values for letters*

A: Students often give a letter a value related to its place in the alphabet, such as b = 2.

C: Students believe that the values of consecutive letters must be consecutive numbers.

O: Students believe that if one letter is before another in the alphabet, its value must be smaller.

R: When the same letter is used more than once in an expression, students wish to state this value separately for each occurrence.

Fig. 4 *Values for letters 2012* – summary of stages and misconceptions

answer any of the 'easy' items for some extraneous reason (strange things sometimes happen in classrooms!) would be flagged as not having met criteria for the first stages, even though they may have answered the most advanced items correctly. A more common example might be a student who is expert at the addition algorithm for fractions and uses it to solve some complex questions (albeit in a complicated way), but who cannot answer apparently more elementary questions about the meaning of fractions.

2.3 Reporting Developmental Stages and Misconceptions

> [Evidence generated to support learning needs to be] *more than information about the presence of a gap between current and desired performance. The evidence must also provide information about what kinds of instructional activities are likely to result in improving performance.* (Wiliam, 2011, p. 11)

A major design decision has been how to present results to teachers. We wanted to use computing power to move away from using behavioural item-by-item descriptions (i.e., saying what items student get correct) to look at broader stages of conceptual development that give teachers more insight into student thinking. We also wanted to help teachers understand how students perceive mathematical ideas. The approach that we selected is to describe learning in terms of topic-specific stages along a learning hierarchy. Our website calls these 'developmental stages.' We decided to report on each student's stage in the specific topic and also flag if they exhibited any misconceptions or common errors (Stacey, Price, & Steinle, 2012).

A learning hierarchy is created by considering some combination of the following: postulating a complexity order based on logical analysis; using teaching experience; using prior research; and analysing empirical data. Stages in a learning hierarchy are confirmed by data if several conditions are met. We require items with similar mathematical characteristics to have similar success rates, and to be completed successfully by the same students. If these conditions are not met, the items need further investigation. When a learning hierarchy exists, knowledge at one stage is prerequisite for achieving tasks at a higher stage. This means that students unable to complete items designed to test lower stages will be less likely than other students to successfully complete items designed to test higher stages. There are many complexities in this simplified story: for example, some of our learning hierarchies have branches. A detailed example is given by Stacey, Price and Steinle (2012).

Figure 4 shows the four developmental stages and the misconceptions and common errors created for the 2012 *Values for letters* smart-test described above. This is a later version of the test and developmental stages than reported by Steinle et al. (2009) because analysis of more student data caused us to add more items, and so improve the reliability and range of diagnosis. More items improve the reliability of diagnosis because there is more capacity to discount the influence of careless (i.e., unsystematic) errors and hence more capacity to be sure that patterns in responses are a true reflection of students' thinking. The new items explore alphabetic misconceptions and they check that students have the basic knowledge of substitution to complete the items meaningfully. The developmental stages go from early use of letters as a code for numbers, to basic understanding of letters as a 'placeholder' for a number, and on to refined understanding. Three of the misconceptions relate to alphabetical interpretations of algebraic letters that can linger to influence students' thinking at various levels of competence, and the fourth is an interesting error that reveals lingering uncertainty about letters as placeholder (students rejecting '$x = 4$' in the item above but accepting '$x = 4$, $x = 4$, $x = 4$').

These stages are empirically confirmed, although this does not mean that every student 'goes through' each of these stages. Well taught students may, for example, very quickly jump from Stage 1 (a basic idea of a letter as a placeholder for a number) to Stage 4. In fact, this test is unusual in that the developmental stages might also be seen as identifying misconceptions and a somewhat arbitrary (but empirically confirmed) decision has been made to put one sequence of misconceptions into the stages and flag others separately. This is due to the very specific nature of this test. A broader test (see, e.g., Stacey et al., 2012) uses larger steps of new knowledge for the basis of the stages.

There are many design decisions related to the presentation of results so that their usefulness to teachers is maximised. In order for the smart-test system to improve teachers' pedagogical content knowledge, information needs to be accurate, complete, and well-researched, but busy teachers are unlikely to spend considerable time reading a large amount of text. Space precludes discussion of most of these design issues. Relevant to this paper is the major issue of finding the appropriate level of detail and of technical language to describe stages. To this end, we have prominent brief versions (as in Fig. 4) hyperlinked to detailed explanations and examples. Similar considerations apply to the teaching advice which is given for students at each stage.

3 Helping Teachers Transform Diagnostic Information into Formative Assessment

The sections above have presented smart-tests from the point of view of the designers. As well as analysing student data to improve the items and diagnoses, we have sought feedback (see Sect. 1.1) from teachers throughout the development of the system. As a result, many additional features have been included, a few of which have been mentioned above, and we have also developed advice for teachers and school leaders to use smart-tests effectively. In essence, this section demonstrates that, while the smart-test system takes over some aspects of diagnostic competence, other aspects of diagnostic competence are needed to use the tool effectively. Because smart-tests are different from the tests that teachers normally set and students take, teachers are likely to initially experience some disequilibrium. Here, we explore two frequently raised issues which are related to this disequilibrium: teachers' assumptions about the purpose of the assessment (summative or formative); and their request for the system to provide feedback directly to students.

3.1 Appreciating Formative Assessment

We found that the concept of formative assessment – assessment that directly feeds into lesson planning – was not understood by all teachers, even in the three schools which had volunteered to trial our first diagnostic tests. During the first 2 years of development, we visited the participating teachers and schools several times each year to get feedback from teachers and conducted formal focus groups. One of the questions regularly raised by the teachers was how to use the smart-test system information in their biannual formal reports to parents. These group discussions showed us that teachers primarily wanted summative rather than formative assessment. Teachers have many demands made upon their time and some of the teachers in the participating schools were hoping to use the smart-test system directly in the time-consuming task of writing reports. Some teachers also expected a measurement of the overall level that students had reached in mathematics against the published state standards both for reporting to parents and for accountability to the local department of education. At that time, an increase of one level in the published state standards indicated 6 months of average mathematical growth.

An early issue arising was therefore whether we should adapt our assessment system to meet teachers' desire for summative assessment. We rejected this proposal for various reasons, and retained the focus on diagnosis to inform immediate teaching. For example, there is no easy marriage between global reporting levels and our stages – our developmental stages remain 'local', referring only to the topic. Stage 1 in one topic has no particular relation to Stage 1 in any other. Stages are also not related to the global achievement levels of any official curriculum documents, even though at first glance this might make summative reporting much easier for teachers. One reason for this disparity in stages is their different 'grain size'. Sometimes, diagnostic developmental stages are very fine grained – we intend that students can move through very quickly with appropriate instruction, possibly in a couple of lessons. The stages for the test *Values for letters* in Fig. 4 are like this. On the other hand, we know that for some tests where complex concepts are involved, students on average take some years to move through the stages. For example, one of our smart-tests, described in Baratta, Price, Stacey, Steinle, and Gvozdenko (2010), looks at percentage problems with different quantities unknown and numbers of varying complexity, and in this case the stages take some years for many students to master. We saw that many teachers took quite some time to broaden their understanding of useful assessment to include formative as well as summative purposes and to appreciate that data on students could be used in the short-term to match instruction to students' needs. One very successful approach to building up an appreciation of formative assessment (confirmed by internet traffic of use and reuse of smart-tests) was including smart-tests in the professional development organised by the Project Officer. She worked with about 50 schools in 2009–2015, and visited teachers to assist with implementation. In the programmes, teachers gathered data from their own classes through the smart-test system and used the diagnoses in their lesson planning over an extended period. Results from particular smart-tests were also used in the face-to-face sessions to show teachers the relevance to their own teaching.

The Project Officer, when first interviewed in 2012, articulated two common views that she had encountered during early professional development sessions (see Sect. 1.1) which she felt hindered teachers' uptake of formative assessment. The first view was that teachers already know the students in their classes and so can accommodate their needs without any additional specific information. These teachers observe the general mathematical achievement of their students and often plan lessons or goals for broad groupings of high-, middle-, and low-achieving students. They do not, however, consider the particular understanding that each student or groups of students have, in different mathematical topics. Instead, they tend to expect students to master simple, medium, or complex aspects of a topic. The second view is that in every class there is a variety of students, so the whole scope of each topic needs to be covered with the whole class. Therefore, there is no need to identify individual student understanding as every student will do every aspect of the topic at the expected level of difficulty. The Project Officer commented in 2012 that both of these views were prevalent among teachers and enabled them to teach in the same way each year with little or no differentiation for differences between classes or individuals.

In the interview 3 years later, she noted that there has been movement in the use of assessment in schools in the education system in which she conducts the professional development programmes, and some shift in the views expressed by teachers. The teachers have been encouraged by the local education authorities to change their view of what 'knowing your students' actually means. She reported that she had not heard teachers saying that the whole class should go through the whole scope of the topic for several years. There is also some evidence from another study (Quenette, 2014) that the views that some teachers hold about their students being either 'good' or 'bad' at mathematics are ameliorated when smart-tests help them appreciate that students who are not making progress may be held up by a gap in knowledge or a misconception, and that even capable students can have misconceptions in some areas.

During the 2012 interview, the Project Officer reported that the turning point for the appreciation of formative assessment for some of her professional development participants came when they realised that formative assessment could sometimes save teaching time, by increasing efficiency of learning. She also noted the importance of the first practical use of a smart-test being undertaken within the supportive professional development environment. Examining the results helped teachers realise that their assumptions about student understanding and competence were not always accurate, and that some students who are procedurally competent have misconceptions. Some teachers realised for the first time that they might accidentally write a test on which students with misconceptions could score 100% if they do not use an appropriately wide range of items types (including non-prototypical) and items which probe thinking. Procedural fluency often masks lack of understanding and the latter can hinder long term progress.

For teachers not in a professional development programme, the smart-test system provides some information about the use of formative assessment when teachers sign up to use the system. The information stresses its special and different character and gives examples of how it can be used.

3.2 *The Provision of Feedback Directly to Students*

Another issue that many teachers raised early in the project is whether the automated diagnosis should be delivered to teachers only or also directly to students. There are two drivers of this request. Firstly, most computer games or quizzes that students use provide immediate feedback (usually right/wrong), so students expected this from online tests. Secondly, many teachers are aware that good quality feedback, presented to students soon after the completion of a task, can lead to increases in learning. In fact, involving students in the results of assessment is often cited as a hallmark of good formative assessment (see, for example, Sadler, 1989).

After consideration, we have maintained our position to provide information only to teachers. We want teachers at the centre of the diagnostic process, because we believe that substantial teacher input is required to overcome most of the conceptual obstacles identified. Furthermore, the detailed topic-specific diagnoses are written for adults, and some effort, background and technical language is required to understand them. Student feedback would need to be written separately and at a variety of reading levels. Leader 1 was interviewed about a range of implementation issues in her school. In the interview, she described how teachers adapt the smart-test feedback they receive to describe in age-appropriate terms for students what the student has shown he or she can do and also the stage toward to which he or she will be working.

Another issue in potentially providing diagnoses direct to students was the disagreement among teachers on the nature of the feedback that might go directly to students. For example one group of teachers warned about negative consequences of students receiving feedback which indicated low performance and hence they recommended only good performances to be reported to students.

We have now resolved the student feedback dilemma by advising teachers to explain the purpose of formative assessment to students before the test, so students know that the information gathered will be used to their advantage, whether as individual feedback or to adapt class lessons. Feedback from teachers, such as Teacher A in Fig. 5, indicates that this is a successful strategy. The observations by the Project Officer (see Fig. 5) confirm this general impression. As a separate project, we are currently developing a modified system which provides diagnostic information direct to test-takers who are pre-service teachers.

4 Building Diagnostic Competence and Improving Teaching

Our aim for the smart-tests is to improve student learning and also to improve teachers' knowledge for teaching (especially their pedagogical content knowledge). In this section, we report on teachers' views about the effect of using the smart-test system on their knowledge for teaching and whether (as well as how) it has changed their practice. Achieving these two goals will really make building the smart-test system worthwhile.

Teacher A: "I just explain to the kids what it shows, and that it's showing me how to teach […] better. 'It's not about something that you're going to get tests back. It's just a tool that I'm using to see what you guys know, so I can teach you better.' They have had no issues with that. And the parents that I've spoken to on parent teacher night a couple of times think it is fantastic."

Project Officer: "Teachers explained the purpose of the assessment to students and … the students became relaxed when the teacher followed through with the intent."

Fig. 5 Comments related to providing assessment results to students

4.1 Effect on Knowledge for Teaching

As is shown in Fig. 1, one of the aims of the smart-test system is to increase teachers' mathematical pedagogical content knowledge. We hypothesised that putting data on their own students' thinking into teachers' hands would make research results come alive for teachers, and hence build their capacity to teach both current and future students. In this sense, the smart-test system is trying to take research results out of the library and put them into the hands and minds of teachers.

In his 2015 interview, Leader 2 expressed the opinion that he found smart-test diagnostic information useful for teacher learning:

> It is difficult to find time to lift the mathematics content knowledge of the teachers. …. This is an ideal way of me being able to introduce a little bit of professional development informally… I feel easier because I know that the teachers are getting some professional development.

Some data on whether teachers feel their knowledge base has improved comes from the voluntary survey, completed by teachers after they accessed students' results from a test. The survey included multiple choice items, with space for optional comments on each item. Table 1 provides the frequency of survey responses to the multiple choice question: *As a result of using this quiz, have you learned something useful for you as a teacher?* The results show that nearly all respondents reported that they learned something useful and nearly half chose 'very valuable learning'.

The associated comments provide further evidence that teachers perceived that using the smart-test system has led to improvement in their knowledge for teaching. Sample comments are provided in Fig. 6. Both teachers B and C reported their own increased confidence in understanding how students think, whilst teachers D, E and F added successful new teaching strategies to their repertoires from the teaching advice provided by the system.

We acknowledge that self-reporting has limitations as a method of data collection, both because of the volunteer sample and in the opinions expressed, so we are cautious in the use of the data obtained. We expect those with strong opinions for or against the smart-tests to be over-represented. Since there was no pressure on teachers to make positive comments about smart-tests (and indeed our questions encouraged specific suggestions for system improvement) we expected that the direction

Table 1 Online survey responses to increasing pedagogical content knowledge

Options provided	Frequency	Percent (%)[a]
YES, very valuable learning	115	47
YES, useful learning	117	47
NO	15	6
Blank	16	–
Total	263	

[a]Percent of 247 non-blank responses

of the comments would indicate the general feeling but that the strength of the opinions offered would be stronger than that of the general teacher population.

The very high proportion of "yes" responses in Table 1 is likely to be due to the fact that many of the teachers were using a particular smart-test for the first time when they completed the surveys. We expect that, on subsequent use of the same test, teachers will be more familiar with the developmental stages, and so they are unlikely to report valuable 'new' learning, except through the results of individual students. In fact, we intend that they will come to observe the developmental stages and misconceptions in their normal interactions with students. In this way, we hope the smart-tests may become redundant, as teachers modify their teaching to reduce the likelihood of misconceptions, help students to develop strong conceptual understanding, and have at their fingertips items which reveal understanding in the specific topic. For example, after knowing about the phenomena revealed by the *Values for letters* test described above, teachers can quite easily address students' false assumptions in their teaching and take care to look for them in students' work. If the test is no longer required because of increased teacher understanding of students' thinking, then that is itself a success.

4.2 Effect on Teaching Practice

An assessment is only formative if it results in a change in the opportunity for a student to learn. Table 2 provides the frequency of responses to the voluntary survey multiple choice question: *Did you adjust your teaching plan as a result of the diagnostic information?* Of the 220 responses to this question, 70% indicated that they did adjust their teaching. Of course, adjusting is not always required. One of the teachers who did not adjust their teaching commented: 'I didn't adjust my teaching plan as such, because the results supported what I expected, but confirmation was valuable'.

The follow-up question to teachers was: *If YES: In what way did you change your teaching plan?* There were many different types of responses. Two very frequent themes are illustrated in Figs. 7 and 8. The first theme, illustrated by comments in Fig. 7, was that teachers used smart-test information to target teaching to specific groups or individuals, especially for overcoming misconceptions or revising basic knowledge.

The second theme concerned changes to the starting point for a unit of work for the whole class. Teachers K, L, M, and N (see Fig. 8), are examples of the many who commented that they started their teaching of the tested topic at a higher level than they had earlier planned. We had expected that many teachers would be alerted to students' inadequate preparation for learning a topic, and so have to start their teaching at a lower level than expected as described by Teacher O, but the comments showed that the opposite situation also commonly occurred.

In the 2012 interview, the Project Officer reported her observation that some teachers had become more centred on the individuals in the class. More often,

Teacher B: "It certainly has encouraged a dialogue between the student and the teacher, and looking at specific things because you as a teacher feel more confident about what you're talking about, because you've got all that information there. The smart-test directs you about where to go. And also you can speak to that student about that particular misconception. It works quite well."

Teacher C: "Well worth doing. Made me feel like an 'expert' teacher instead of just an experienced teacher."

Teacher D: "[I] used a table structure similar to dual number line to help students with showing and organising information contained in problems and to find what 1 part represents [and also] to emphasize the use of multiplication/division."

Teacher E: "I WILL use more materials and a lot more justification from the students." (emphasis used by teacher)

Teacher F: "I read the referenced research paper, which was informative and useful. The teaching suggestions were really practical, and were suitable to have a go at straight away. I used paper strips and pieces to fold and colour to estimate percentages."

Leader 1: "I think teachers are now (since they have been using smart-test data for planning) more confident in … identifying where the students are at."

Fig. 6 Comments related to improvement in knowledge for teaching

Table 2 Online survey responses to teachers adjusting teaching plan

Options provided	Frequency	Percent (%)[a]
YES	154	70
NO	66	30
blank	41	–
Total	261	

[a]Percent of 220 non-blank responses

teachers now planned in more detail for their particular class instead of using the methods that they always used for a particular topic.

Since the smart-test system is used, in the main, by volunteer teachers, and only some of these volunteers spend the time to fill in the survey in a detailed way, it might be expected that general feedback on the system is biased towards the positive. (Negative feedback tends to focus on small technical issues.) Even with this caution, it is good to know that teachers like P, Q, and R in Fig. 9 find the system very helpful.

The comments above show that the smart-tests have had an impact on the teaching of some individual teachers. However, we have observed that for some teachers, data from smart-tests seems to be better used collaboratively to inform changes to

Teacher G: "I often overlook and brush off students' misconceptions without considering the difficulty that students face. With this assessment tool, I am able to analyse my students better individually, and correct their misconceptions on a particular topic."

Teacher H: "I have put the students into groups and will give them activities to focus on and correct their misconceptions. I will be looking carefully at the [suggested resources]."

Teacher I: "Very useful as a pre-test on reading scales. I found out exactly where each student was at and that enabled me to target my teaching into the areas where it was most needed, while giving extension work to the students who had already gained a good understanding of the topic. Now I am going toretest them using another form of the test to see how effective my teaching has been."

Teacher J: "I had assumed that at year 10 my students would have a basic understanding of the idea of percentages - many of them didn't! Instead of going straight into calculating percentages of quantities and calculating whole quantities given a percentage, and then on to financial arithmetic (simple interest), I went back to basics with the students who needed it, and others who could cope with this were assigned the original tasks I had planned."

Fig. 7 Comments related to changes at the level of individuals or groups

teaching, at least at the beginning. The professional development initiatives of the Project Officer, as outlined above, provide evidence of this. A similar process was implemented in their own schools by Leaders 1 and 2 and is happening in a small number of other schools. Collaborative work seems particularly important in primary schools where the majority of teachers are not used to dealing with large volumes of data. Leader 1 reported that primary teachers at her school appreciated help with sorting students into groups based on the data provided. At her school, the administration supported joint unit planning:

> I have a really supportive administration ....and they can definitely see the benefit of it. We do have an hour planning for each year level for maths each week. If we didn't have that it would be really hard to do this.

Leader 2 also said that there had been some 'creative' timetabling at his school to enable teachers to analyse smart-test data together and to plan lessons.

Some data has been collected from teachers working with Leader 1 that may give less subjective information about the effect of smart-test use on the knowledge of teachers. Prior to using smart-tests, these teachers were interviewed about what sort of student difficulties they would expect when teaching various topics from the mathematics curriculum. They were asked how they would explain some key concepts. It is intended that a similar interview be given after a year of smart-test use. The project is not yet complete.

Teacher K: "I adapted the simpler task that we were going to approach in class with something that reflected the students' greater level of understanding."
Teacher L: "I used the smart-test 'Understanding angle' with my year 7 class. In my teaching I adopted an approach that best addressed the needs of the students based upon the diagnostic test. I was able to avoid certain areas that were well understood and concentrate on areas that were not."
Teacher M: "I looked at the course outline. As many of my students were very strong in perimeter, we focused more on area and volume."
Teacher N: "The other end of the spectrum is that I've been more confident in moving kids, not making them go over things. I can see 'alright, this child has a really good understanding of fractions'. I'm not going to … make him (or her) repeat all of those skills so I feel more confident in moving them to something else."
Teacher O: "When our Year 7 students did the fractions smart-tests, we were surprised to find many students were at Stage 0. All these years we've always presumed that they were at a particular level but obviously that's not happening, and so that's changed our curriculum, the way we think about teaching fractions."

Fig. 8 Comments related to changing starting points for teaching

Teacher P: "This quiz is a genuinely useful tool to assist in the differentiation of the curriculum. It is efficient and informative."
Teacher Q: "Excellent formative assessment tool which allowed me as a coach to discuss the various misconceptions and student thinking within a year 8 class. It provided teachers with real data that allowed them to address the misconceptions through their teaching."
Teacher R: "I use the smart-tests as a part of my diagnostic 'toolbox'. They are clear, easy to access and give a quick snapshot of where my students' prior knowledge is developed or underdeveloped. This information influences the activities I implement in class, ensuring that the students are being challenged in Mathematics."

Fig. 9 Comments on usefulness of smart-tests for formative assessment

One important indirect measure of the usefulness of smart-tests is their rate of use. Each year from 2008 to the end of 2015 the usage figures have increased. We also track use and reuse of individual tests. When funding allows, we are planning to investigate whether student performance data has improved in schools where smart-tests are routinely used.

A design project, such as the development of the smart-test system, depends on user feedback, especially for polishing the myriad of features that any system has. The data above has principally been collected by us to improve the system, and to ensure that we are offering some teachers a product which they find valuable. It has been successful for this purpose. Other data has also been collected within the educational systems to ensure that funding the system is a good use of their resources. Research to scientifically investigate whether the system does improve student learning outcomes in general requires a different methodology, including careful examination of how the tests are used within the school, both with teachers and students.

5 Conclusion

The intention of the smart-test system is to take the results of research about students' understanding of particular mathematics topics and to embed it into an intelligent system: a tool holding distributed intelligence which amplifies what teachers can do. This paper has reported the views of the early and current users of the system. In general, the surveys report positively on the tests individually and on the system as a whole. However, the wider experience of creating the smart-test system shows that formative assessment is only beginning to be part of the culture of all schools in our region. Some schools are certainly ready for it, and indeed are now actively using this as a standard part of their planning and teaching. Making formative assessment easier through online tools should promote its use, but it also seems important to have professional development showing its advantages and distinctive features, and to provide teachers with advice on implementation. Finally, data from the surveys provides considerable evidence of a self-reported increase in teachers' pedagogical content knowledge and that teachers are using the information in their subsequent lessons.

The smart-test system is an intelligent tool, which is designed to reduce the work that diagnostic activities require of teachers. By providing carefully designed items, many based on research literature, and automating the diagnosis rubrics, the pedagogical content knowledge required by teachers is also reduced. However, as demonstrated in the chapter, diagnostic competence involves more than this – including understanding the purpose of formative assessment, and having the skills to implement it. In summary, diagnostic competence is still required to use smart-tests well whilst in the other direction, the evidence presented shows that using the smart-tests can itself increase diagnostic competence.

Acknowledgements An early version of this paper (Steinle & Stacey, 2012) was presented at the 12th International Congress on Mathematical Education. The smart-test system was developed at the University of Melbourne by Kaye Stacey, Vicki Steinle, Beth Price and Eugene Gvozdenko, with initial funding from the Australian Research Council Linkage Projects Program LP0882176 in partnership with the Department of Education and Early Childhood Development, Victoria, then further contracts with DEECD and the Catholic Education Office (Melbourne). We thank Ms Valerie Everist, Ms Sara McKee and Mr Terence Griffin for assisting teachers with using this system and for participating in interviews, and all the teachers and students who have contributed to this project to date and continue to do so.

References

Baratta, W., Price, E., Stacey, K., Steinle, V., & Gvozdenko, E. (2010). Percentages: The effect of problem structure, number complexity and calculation format. In L. Sparrow, B. Kissane, & C. Hurst (Eds.), *Proceedings of 33rd Annual Conference of the Mathematics Education Research Group of Australia* (pp. 61–68). Fremantle: MERGA. http://www.merga.net.au/documents/MERGA33_BarattaEtAl.pdf

Fujii, T. (2003). Probing students' understanding of variables through cognitive conflict problems: Is the concept of a variable so difficult for students to understand? In N. Pateman, G. Dougherty, & J. Zilliox (Eds.), *Proceedings of the 27th Annual Conference of the International Group for the Psychology of Mathematics Education* (Vol. 1, pp. 49–65). Hawaii: PME.

Helme, S., & Stacey, K. (2000). Can minimal support for teachers make a difference to students' understanding of decimals? *Mathematics Teacher Education and Development, 2*, 105–120. http://www.smartvic.com/smart/research/index.htm. Accessed 7 Jan 2015.

Küchemann, D. (1981). Algebra. In K. M. Hart, M. L. Brown, D. E. Küchemann, D. Kerslake, G. Ruddock, & M. McCartney (Eds.), *Children's Understanding of Mathematics: 11–16* (pp. 102–119). Oxford: John Murray.

Pea, R. (1987). Practices of distributed intelligence and designs for education. In G. Salomon (Ed.), *Distributed cognitions: Psychological and educational considerations* (pp. 47–87). Cambridge, MA: Cambridge University Press.

Pellegrino, J. W., & Quellmalz, E. S. (2010). Perspectives on the integration of technology and assessment. *Journal of Research on Technology in Education, 43*(2), 119–134.

Quenette, J. (2014). *Diagnostic testing and changes to teaching practice in Year 9 mathematics classes* (Unpublished Master of Education thesis). University of Melbourne. Retrieved September 3, 2015, from https://minerva-access.unimelb.edu.au/handle/11343/43027

Sadler, D. R. (1989). Formative assessment and the design of instructional systems. *Instructional Science, 18*, 119–144.

Stacey, K., & MacGregor, M. (1997). Students' understanding of algebraic notation: 11–15. *Educational Studies in Mathematics, 33*, 1–19.

Stacey, K., Price, B., & Steinle, V. (2012). Identifying stages in a learning hierarchy for use in formative assessment – The example of line graphs. In J. Dindyal, L. P. Cheng, & S. F. Ng (Eds.), *Mathematics education: Expanding horizons* (Proceedings of the 35th Annual Conference of Mathematics Education Group of Australasia, pp. 393–400). Adelaide: MERGA.

Stacey, K., Price, B., Steinle, V., Chick, H., & Gvozdenko, E. (2009). *SMART assessment for learning*. Paper presented at Conference of the International Society for Design and Development in Education, Cairns, Australia, September 28–October 1, 2009. http://www.isdde.org/isdde/cairns/pdf/papers/isdde09_stacey.pdf

Stacey, K., & Wiliam, D. (2013). Technology and assessment in mathematics. In M. A. Clements, A. Bishop, C. Keitel, J. Kilpatrick, & F. Leung (Eds.), *Third international handbook of mathematics education* (pp. 721–752). New York, NY: Springer.

Steinle, V. (2004). *Changes with age in students' misconceptions of decimal numbers* (Unpublished doctoral thesis) University of Melbourne. Retrieved September 3, 2015, from https://minerva-access.unimelb.edu.au/handle/11343/39024

Steinle, V., Gvozdenko, E., Price, B., Stacey, K., & Pierce, R. (2009). Investigating students' numerical misconceptions in algebra. In R. Hunter, B. Bicknell, & T. Burgess (Eds.), *Crossing divides* (Proceedings of the 32nd Annual Conference of the Mathematics Education Research Group of Australasia, Vol. 2, pp. 491–498). Wellington: MERGA.

Steinle, V., & Stacey, K. (2012). Teachers' views of using an on-line, formative assessment system for mathematics. In *Pre-proceedings. 12th International Congress on Mathematical Education Topic Study Group*, 33, 8 July–15 July, 2012 (pp. 6721–6730). Seoul: COEX.

Steinle, V., Stacey, K., & Chambers, D. (2006). *Teaching and learning about decimals* (Version 3.1). Retrieved September 3, 2015, from https://extranet.education.unimelb.edu.au/SME/TNMY/

Stewart, R., Wright, B., & Gould, P. (1998). Kindergarten students' progress in the count me in too project. In C. Kanes, M. Goos, & E. Warren (Eds), *Teaching mathematics in new times* (Proceedings of the 21st Annual Conference of the Mathematics Education Research Association of Australasia, Vol. 2, pp. 556–563). Brisbane: MERGA.

Wiliam, D. (2007). Keeping learning on track: Classroom assessment and the regulation of learning. In F. K. Lester Jr. (Ed.), *Second handbook of mathematics teaching and learning* (pp. 1053–1098). Greenwich, CT: Information Age.

Wiliam, D. (2011). What is assessment for learning? *Studies in Educational Evaluation*, *37*, 3–14.

Index

T. Leuders et al. (eds.), *Diagnostic Competence of Mathematics Teachers*, Mathematics Teacher Education 11, DOI 10.1007/978-3-319-66327-2

Printed by Printforce, the Netherlands